OIL & GAS SIDEKICK

Vital Information Fast

By Dennis Evers

DISCLAIMER

WHILE EVERY EFFORT HAS BEEN MADE TO ENSURE THAT THE INFORMATION IS ACCURATE AT THE TIME OF PUBLICATION, THE AUTHOR IS NOT RESPONSIBLE FOR ANY LOSS, LIABILITY, DAMAGE OR INJURY THAT MAY BE SUFFERED OR INCURRED BY ANY PERSON IN CONNECTION WITH THE INFORMATION CONTAINED IN THIS BOOK, OR BY ANYONE WHO RECEIVES FIRST AID TREATMENT FROM A READER OR USER OF THIS INFORMATION.

THIS INFORMATION IS ADVISORY IN NATURE AND IS NOT INTENDED TO IDENTIFY ALL SCENARIOS OR SITUATIONS A PERSON MIGHT FIND THEMSELVES IN.

FOLLOWING THESE GUIDELINES WILL NOT GUARANTEE YOUR SAFETY IN ANY SITUATION.

For quantity pricing go to: sidekick.deellc.com

For Suggestions/Corrections/Custom Publishing
contact dennis@deellc.com

Cover Graphics by sparrowsgift.com
Art by B.J. Evers
Thanks to Dr. Berto Silva

ISBN-10:0692210687

ISBN-13: 978-0692210680

DEDICATION

Oil & gas workers keep the world running, working around the clock in some of the most inhospitable locations on this planet. Drilling holes thousands of feet below terra firma, and from little platforms suspended over miles of deep blue sea, working long, hard hours, producing the fluids that sustain life and create a standard of living that has never been seen before.

The job requires leaving loved ones, spouses and children, often for weeks or months at a time, living in temporary quarters, travelling to jobs locally and halfway around the world, getting the job done with skill, precision and efficiency.

In addition to the hazards on site, many contend with civil unrest, terrorism, desert heat, frozen wasteland and extreme environmentalists, all of whom rely on energy producers in one way or another.

The work is an inherently dangerous profession, but it is constantly being made as safe as possible.

This book is dedicated to the countless professionals' upstream, midstream and downstream who keep the world running. Without you, we would be back in the 1800's.

Godspeed,

Dennis Evers

Table of Contents

COMMON OIL & GAS CHEMICALS

Identification

First Aid

Emergency Procedures

Note: To the best of my knowledge, the information contained herein is accurate. However, this publisher assumes no liability whatsoever for the accuracy or completeness of the information contained herein. Final determination of suitability of any material is the sole responsibility of the user. All materials may present unknown hazards and should be used with caution. Although certain hazards are described herein, I cannot guarantee that these are the only hazards that exist. Not all events are survivable.

For Transportation Emergencies:

CHEMTREC®, Inside the USA:
800- 424-9300

CHEMTREC®, Outside the USA:
703- 527-3887

(CANUTEC- Canada)
1-613-996-6666

POISON CONTROL CENTER
1-800-222-1222.

Emergency Company Numbers:

____

____

____

____

1,1,1 - Trichloroethane (CAS#71-55-6)

COLOR: colorless PHYSICAL FORM: liquid ODOR: sweet odor

NFPA: HEALTH=2 FIRE=1 REACTIVITY=0

Emergency Overview
Respiratory tract irritation, skin irritation, eye irritation, central nervous
system depression.

FIRST AID
INHALATION: If adverse effects occur, remove to uncontaminated area. Give artificial respiration if not breathing. If breathing is difficult, oxygen should be administered by qualified personnel. Get immediate medical attention.
SKIN CONTACT: Wash skin with soap and water for at least 15 minutes while removing contaminated clothing and shoes. Get medical attention, if needed. Thoroughly clean and dry contaminated clothing and shoes before reuse.
EYE CONTACT: Flush eyes with plenty of water for at least 15 minutes. Then get immediate medical attention.
INGESTION: If vomiting occurs, keep head lower than hips to help prevent aspiration. If person is unconscious, turn head to side. Get medical attention immediately.

Protective Equipment and Precautions for Firefighters
As in any fire, wear self-contained breathing apparatus pressure-demand, MSHA/NIOSH (approved or equivalent) and full protective gear.

EXTINGUISHING MEDIA: carbon dioxide, regular dry chemical

4-4' Methylene dianiline (CAS#101-77-9)

Appearance Off-white, Chips Physical State Solid

NFPA Health 2 Flammability 1 Instability 0

Emergency Overview
Toxic. Harmful by inhalation, in contact with skin and if swallowed. Irritant. May cause cyanosis.

FIRST AID
Eye Contact Rinse immediately with plenty of water, also under the eyelids, for at least 15 minutes. Call a physician immediately.
Skin Contact Wash off immediately with soap and plenty of water removing all contaminated clothes and shoes. Call a physician immediately.
Inhalation Move to fresh air. If breathing is difficult, give oxygen. If not breathing, give artificial respiration. Call a physician immediately.
Ingestion Do not induce vomiting. Never give anything by mouth to an unconscious person. Call a physician immediately.

Protective Equipment and Precautions for Firefighters
As in any fire, wear self-contained breathing apparatus pressure-demand, MSHA/NIOSH (approved or equivalent) and full protective gear.

Suitable Extinguishing Media Carbon dioxide (CO2). Dry powder. Foam.

Acetic acid (CAS#64-19-7)

Physical State Liquid   Appearance Colorless   Odor vinegar-like

NFPA Health 3   Flammability 2   Instability 0

Emergency Overview
Flammable liquid and vapor. Causes severe burns by all exposure routes.

FIRST AID
Eye Contact Rinse immediately with plenty of water, also under the eyelids, for at least 15 minutes. Immediate medical attention is required.
Skin Contact Wash off immediately with plenty of water for at least 15 minutes. Immediate medical attention is required.
Inhalation Move to fresh air. If breathing is difficult, give oxygen. Do not use mouth-to-mouth resuscitation if victim ingested or inhaled the substance; induce artificial respiration with a respiratory medical device. Immediate medical attention is required.
Ingestion Do not induce vomiting. Call a physician or Poison Control Center immediately.

Protective Equipment and Precautions for Firefighters

As in any fire, wear self-contained breathing apparatus pressure-demand, MSHA/NIOSH (approved or equivalent) and full protective gear.

Suitable Extinguishing Media Use water spray, alcohol-resistant foam, dry chemical or carbon dioxide.

Acetic anhydride (CAS#108-24-7)

Physical State Liquid Appearance Colorless Odor Pungent

NFPA Health 3 Flammability 2 Instability 1

Emergency overview
Flammable liquid and vapor. May be fatal if inhaled. Exposure through inhalation may result in delayed pulmonary edema, which may be fatal. Harmful if swallowed. Causes severe eye and skin burns. Lachrymator (substance which increases the flow of tears). Reacts violently with water.

FIRST AID
Eye Contact Rinse immediately with plenty of water, also under the eyelids, for at least 15 minutes. Immediate medical attention is required.
Skin Contact Wash off immediately with plenty of water for at least 15 minutes. Immediate medical attention is required.
Inhalation Move to fresh air. If breathing is difficult, give oxygen. Do not use mouth-to-mouth resuscitation if victim ingested or inhaled the substance; induce artificial respiration with a respiratory medical device. Immediate medical attention is required.
Ingestion Do not induce vomiting. Call a physician or Poison Control Center immediately.

Protective Equipment and Precautions for Firefighters
As in any fire, wear self-contained breathing apparatus pressure-demand, MSHA/NIOSH (approved or equivalent) and full protective gear.

Suitable Extinguishing Media Carbon dioxide (CO2). Dry chemical. chemical foam. Flooding quantities of water. Cool closed containers exposed to fire with water spray.

Acetone (CAS#67-64-1)

Clear, colorless liquid Sweet mint-like odor

HMIS Ratings: Health: 2 Fire: 3 Physical Hazard: 0

EMERGENCY OVERVIEW: This product is a clear, volatile, flammable liquid. Has a sweet, mint-like odor. Highly flammable.

SKIN: Wash off immediately with soap and plenty of water. Take off contaminated clothing and shoes immediately. Wash contaminated clothing before re-use. Obtain medical attention.
EYES: Rinse immediately with plenty of water, also under the eyelids, for at least 15 minutes. Obtain medical attention.
INHALATION: Move to fresh air in case of accidental inhalation of vapours. If not breathing, give artificial respiration. If breathing is difficult, give oxygen, provided a qualified operator is available. Call a physician immediately.
INGESTION: DO NOT induce vomiting. Immediate medical attention is required. If vomiting occurs naturally, have victim lean forward to reduce risk of aspiration.

Protective Equipment and Precautions for Firefighters
As in any fire, wear self-contained breathing apparatus pressure-demand, MSHA/NIOSH (approved or equivalent) and full protective gear.

EXTINGUISHING MEDIA:
Use alcohol-resistant foam, carbon dioxide (CO2) or dry chemical.
UNUSUAL FIRE AND EXPLOSION HAZARDS:
Highly flammable. Vapours may form explosive mixtures with air. Vapours are heavier than air and may travel along the ground to some distant source of ignition and flash back.

Acetylene gas (CAS#74-86-2)

NFPA Health:0 Flammability:4 Instability:3

Emergency overview
FLAMMABLE GAS. MAY CAUSE FLASH FIRE. CONTENTS UNDER PRESSURE.
No action shall be taken involving any personal risk or without suitable training. If it is suspected that fumes are still present, the rescuer should wear an appropriate mask or self-contained breathing apparatus. It may be dangerous to the person providing aid to give mouth-to-mouth resuscitation.

FIRST AID:

EYES: Check for and remove any contact lenses. Immediately flush eyes with plenty of water for at least 15 minutes, occasionally lifting the upper and lower eyelids. Get medical attention immediately.
SKIN: In case of contact, immediately flush skin with plenty of water for at least 15 minutes while removing contaminated clothing and shoes. To avoid the risk of static discharges and gas ignition, soak contaminated clothing thoroughly with water before removing it. Wash clothing before reuse. Clean shoes thoroughly before reuse. Get medical attention immediately
Frostbite: Try to warm up the frozen tissues and seek medical attention.
Inhalation: Move exposed person to fresh air. If not breathing, if breathing is irregular or if respiratory arrest occurs, provide artificial respiration or oxygen by trained personnel. Loosen tight clothing such as a collar, tie, belt or waistband. Get medical attention immediately.

Protective Equipment and Precautions for Firefighters
As in any fire, wear self-contained breathing apparatus pressure-demand, MSHA/NIOSH (approved or equivalent) and full protective gear.

Fire-fighting:In case of fire, use water spray (fog), foam or dry chemical.
Allow gas to burn if flow cannot be shut off immediately. Apply water from a safe distance to cool container and protect surrounding area. If involved in fire, shut off flow immediately if it can be done without risk. Contains gas under pressure. Flammable gas. In a fire or if heated, a pressure increase will occur and the container may burst, with the risk of a subsequent explosion.

Acrylamide Monomer (CAS#79-06-1)

Physical State: Solid   Appearance: white   Odor: Odorless.

NFPA Rating: (estimated) Health: 2; Flammability: 2; Instability: 2

Emergency overview
Acrylamide may cause nervous system damage. Acrylamide may form explosive dust-air mixtures. Harmful if swallowed, inhaled, or absorbed through the skin. Causes eye irritation. May cause allergic skin reaction.

FIRST AID
Eyes: In case of contact, immediately flush eyes with plenty of water for a t least 15 minutes. Get medical aid.
Skin: In case of contact, immediately flush skin with plenty of water for at least 15 minutes while removing contaminated clothing and shoes. Get medical aid immediately. Wash clothing before reuse.
Ingestion: If swallowed, do not induce vomiting unless directed to do so by medical personnel. Never give anything by mouth to an unconscious person. Get medical aid.
Inhalation: If inhaled, remove to fresh air. If not breathing, give artificial respiration. If breathing is difficult, give oxygen. Get medical aid.

Protective Equipment and Precautions for Firefighters
As in any fire, wear self-contained breathing apparatus pressure-demand, MSHA/NIOSH (approved or equivalent) and full protective gear.

Extinguishing Media: Use water spray, dry chemical, carbon dioxide, or chemical foam

Acrolein (CAS#107-02-8)

COLOR: colorless to yellow PHYSICAL FORM: volatile liquid
ODOR: pungent odor

NFPA RATINGS : HEALTH=4 FIRE=3 REACTIVITY=3

EMERGENCY OVERVIEW
May explode when heated. Flammable liquid and vapor. Vapor may cause flash fire
SHORT TERM EXPOSURE: irritation (possibly severe), tearing, nausea, vomiting, diarrhea, difficulty breathing, asthma, headache, drowsiness, symptoms of drunkenness, fainting, bluish skin color, lung damage, death

FIRST AID
INHALATION: If adverse effects occur, remove to uncontaminated area. Give artificial respiration if not breathing. If breathing is difficult, oxygen should be administered by qualified personnel. Get immediate medical attention.
SKIN CONTACT: Wash skin with soap and water for at least 15 minutes while removing contaminated clothing and shoes. Get immediate medical attention. Thoroughly clean and dry contaminated clothing before reuse. Destroy contaminated shoes.

EYE CONTACT: Immediately flush eyes with plenty of water for at least 15 minutes. Then get immediate medical attention.
INGESTION: Contact local poison control center or physician immediately. Never make an unconscious person vomit or drink fluids. When vomiting occurs, keep head lower than hips to help prevent aspiration. If person is unconscious, turn head to side. Get medical attention immediately.

Protective Equipment and Precautions for Firefighters
As in any fire, wear self-contained breathing apparatus pressure-demand, MSHA/NIOSH (approved or equivalent) and full protective gear.

FIRE AND EXPLOSION HAZARDS: Severe fire hazard. Vapor/air mixtures are explosive. The vapor is heavier than air. Vapors or gases may ignite at distant ignition sources and flash back.

EXTINGUISHING MEDIA: regular dry chemical, carbon dioxide, water, regular foam, alcohol-resistant foam
Large fires: Use regular foam or flood with fine water spray

Aluminum chloride (CAS#7446-70-0)

Physical State - Crystalline solid or powder. Color - Gray yellow and white particles. Odor - Sharp, pungent, and acidic

NFPA Hazard Health: 3 Flammability: 0 Reactivity: 2 W

EMERGENCY OVERVIEW
Causes burns.Hygroscopic (absorbs moisture from the air).Moisture sensitive.

FIRST AID
Eyes: Flush eyes thoroughly with large quantities of water for at least 15
minutes. Seek medical attention immediately.
Skin: Brush off any solid aluminum chloride before washing with soap and
water or thermal burns will result from the reaction with water. When flushing with water, use large amounts of water. If irritation or burns develop seek immediate medical attention.
Inhalation: Remove victim to fresh air. If breathing is difficult, administer

oxygen. If not breathing, perform rescue breathing. Seek immediate medical attention. Aluminum chloride reacts with water to form hydrochloric acid which can be corrosive to the throat and lungs. Treatment should be as appropriate for chemical or thermal burns to the lungs.
Ingestion: Seek immediate medical attention. Treatment should be as
appropriate for acids. Aluminum chloride reacts with water in the system to form
hydrochloric acid. Chemical and thermal burns to the mouth, throat and stomach may occur. Treat as appropriate for acid burns or thermal burns.

Protection of First-Aiders: If it is suspected that fumes are still present, the rescuer should wear an appropriate mask or self-contained breathing apparatus. It may be dangerous to the person providing aid to give mouth-to-mount resuscitation. Wash contaminated clothing thoroughly with water before removing it, or wear gloves.

Protective Equipment and Precautions for Firefighters
As in any fire, wear self-contained breathing apparatus pressure-demand, MSHA/NIOSH (approved or equivalent) and full protective gear.

Fire fighting instructions. Do not put water on aluminum chloride spills! Water reacts violently and exothermically with aluminum chloride releasing toxic and corrosive hydrogen chloride gas. If water gets into closed containers or vessels the vessels could rupture do to over-pressurization. Firefighters should do everything possible to keep water or moisture away from Aluminum Chloride. Utilize SCBA's and full turnout gear to respond. If vapors are in high concentrations (cause irritation to the skin), fully encapsulated Level A Suits are required.

Extinguishing media Use an extinguishing media other than water which is suitable for a surrounding fire. Do not use water to extinguish surrounding fire as a violent
exothermic reaction producing corrosive hydrogen chloride gas can result.

Ammonium bifluoride (CAS#1341-49-7)

Appearance: White crystals. Odor: Odorless.

NFPA Ratings: Health: 3 Flammability: 0 Reactivity: 0

Emergency Overview
DANGER! MAY BE FATAL IF SWALLOWED OR INHALED. AFFECTS RESPIRATORY SYSTEM, HEART, SKELETON, CIRCULATORY SYSTEM, CENTRAL NERVOUS SYSTEM AND KIDNEYS. CAUSES IRRITATION AND BURNS TO SKIN, EYES AND RESPIRATORY TRACT. IRRITATION AND BURN EFFECTS MAY BE DELAYED. HARMFUL IF ABSORBED THROUGH SKIN

First Aid First aid procedures should be **pre-planned** for fluoride compound emergencies.
Inhalation: If inhaled, remove to fresh air. If not breathing, give artificial respiration. If breathing is difficult, give oxygen. CALL A PHYSICIAN IMMEDIATELY.
Ingestion: Administer milk, chewable calcium carbonate tablets or milk of magnesia. Never give anything by mouth to an unconscious person. CALL A PHYSICIAN IMMEDIATELY.
Skin Contact: Wipe off any excess material from skin and then immediately flush skin with large amounts of soapy water. Remove contaminated clothing and shoes. Wash clothing before re-use. Apply bandages soaked in magnesium sulfate. CALL A PHYSICIAN IMMEDIATELY.
Eye Contact: Immediately flush eyes with gentle but large stream of water for at least 15 minutes, lifting lower and upper eyelids occasionally. Call a physician immediately.

Protective Equipment and Precautions for Firefighters
As in any fire, wear self-contained breathing apparatus pressure-demand, MSHA/NIOSH (approved or equivalent) and full protective gear.

Fire: Not considered to be a fire hazard.
Explosion: Contact with water and metal at the same time may evolve flammable hydrogen gas.
Fire Extinguishing Media: Dry chemical, foam or carbon dioxide. Do not use water.

Ammonium bisulfite (CAS#10192-30-0)

Appearance: Colorless to pale yellow liquid. Odor: Slight sulfur dioxide odor.

NFPA: Heath - 1 Flammability - 0 Reactivity - 1

Emergency Information:

Solutions are slightly acidic (pH 5.5) Contact may cause eye irritation. Repeated/prolonged skin contact may cause irritation. Avoid inhalation of fumes in vapor space (sulfur dioxide). Ingestion may irritate gastrointenstlnal tract. Heating may cause ammonia and sulfur dioxide gas to evolve.

FIRST AID

EYES: Immediately flush with large quantities of water for 15 minutes. Hold
eyelids apart during irrigation to insure thorough flushing of the entire area
of the eye and lids. Obtain immediate medical attention.

SKIN: Immediately flush with large quantities of water. Remove contaminated
clothing under a safety shower. Obtain immediate medical attention.

INGESTION: If victim is conscious, immediately give 2 to 4 glasses of water.
Induce vomiting by touching finger to back of throat. Obtain immediate
medical attention.

INHALATION: Remove victim from contaminated atmosphere. If breathing is
labored, administer oxygen. If breathing has ceased, clear airway and start
mouth to mouth resuscitation. If heart has stopped beating, external heart
massage should be applied. Obtain immediate medical attention.

Protective Equipment and Precautions for Firefighters

As in any fire, wear self-contained breathing apparatus pressure-demand, MSHA/NIOSH (approved or equivalent) and full protective gear.

Small Releases: Confine and absorb small releases on sand, earth or other inert
absorbent. Dispose of in a chemical waste landfill.

Large Releases: Confine area to qualified personnel. Wear proper protective
equipment. Shut off release if safe to do so. Dike spill area to prevent runoff
into sewers, drains or surface waterways (potential aquatic toxicity). Spray
product vapors with water spray or mist. Recover as much of the solution as
possible. Treat remaining material as a small release (above).

Ammonium fluoride (CAS#12125-01-8)

Color and Form white crystal. Odor: Ammonia-like

HMIS ratings Health (acute effects) = 2 Flammability = 0 Reactivity = 1

Emergency Overview:
Toxic by inhalation, in contact with skin and if swallowed. This compound may cause fluoride poisoning by reducing cellular calcium function. Early symptoms include nausea, vomiting, diarrhea, and weakness.

FIRST AID
Eye Exposure: Immediately flush the eyes with copious amounts of water for at least 10-15 minutes. A victim may need assistance in keeping their eye lids open. Get immediate medical attention.
Skin Exposure: Wash the affected area with water. Remove contaminated clothes if necessary. Seek medical assistance if irritation persists.
Inhalation: Remove the victim to fresh air. Closely monitor the victim for signs of respiratory problems, such as difficulty in breathing, coughing, wheezing, or pain. In such cases seek immediate medical assistance.
Ingestion: Seek medical attention immediately. Keep the victim calm. Do not induce vomiting.

Protective Equipment and Precautions for Firefighters
As in any fire, wear self-contained breathing apparatus pressure-demand, MSHA/NIOSH (approved or equivalent) and full protective gear.

Suitable extinguishing agents Carbon dioxide, extinguishing powder or water spray. Fight larger fires with water spray or alcohol resistant foam.

Ammonium persulfate (CAS#7727-54-0)

Appearance: white to light yellow powder.

NFPA: (estimated) Health: 2; Flammability: 1; Instability: 2; Special Hazard: OX

EMERGENCY OVERVIEW

Strong oxidizer. Contact with other material may cause a fire. Causes eye, skin, and respiratory tract irritation. May cause allergic respiratory and skin reaction. Harmful if swallowed. May cause sensitization by inhalation and by skin contact.

FIRST AID

Eyes: Immediately flush eyes with plenty of water for at least 15 minutes, occasionally lifting the upper and lower eyelids. Get medical aid.

Skin: Get medical aid immediately. Immediately flush skin with plenty of water for at least 15 minutes while removing contaminated clothing and shoes.

Ingestion: Get medical aid. Do NOT induce vomiting. If conscious and alert, rinse mouth and drink 2-4 cupfuls of milk or water.

Inhalation: Get medical aid immediately. Remove from exposure and move to fresh air immediately. If breathing is difficult, give oxygen. Do NOT use mouth-to-mouth resuscitation. If breathing has ceased apply artificial respiration using oxygen and a suitable mechanical device such as a bag and a mask.

Protective Equipment and Precautions for Firefighters

As in any fire, wear self-contained breathing apparatus pressure-demand, MSHA/NIOSH (approved or equivalent) and full protective gear.

Extinguishing Media: Use water spray, dry chemical, or carbon dioxide. Cool containers with flooding quantities of water until well after fire is out.

Ammonium chloride (CAS#12125-02-9)

Physical State: Solid Appearance: colorless or white
Odor: odorless

NFPA Rating: (estimated) Health: 1; Flammability: 0; Reactivity: 0

EMERGENCY OVERVIEW

May cause respiratory and digestive tract irritation. May be harmful if swallowed. Causes eye irritation. May cause skin irritation.

FIRST AID

Eyes: Flush eyes with plenty of water for at least 15 minutes, occasionally lifting the upper and lower eyelids. Get medical aid immediately.
Skin: Flush skin with plenty of soap and water for at least 15 minutes while removing contaminated clothing and shoes. Get medical aid if irritation develops or persists. Wash clothing before reuse.
Ingestion: Induce vomiting. If victim is conscious and alert, give 2-4 cupfuls of milk or water. Never give anything by mouth to an unconscious person. Get medical aid.
Inhalation: Remove from exposure to fresh air immediately. If not breathing, give artificial respiration. If breathing is difficult, give oxygen. Get medical aid.

Protective Equipment and Precautions for Firefighters
As in any fire, wear self-contained breathing apparatus pressure-demand, MSHA/NIOSH (approved or equivalent) and full protective gear.

Extinguishing Media: For small fires, use dry chemical, carbon dioxide, water spray or alcohol-resistant foam. Substance is noncombustible; use agent most appropriate to extinguish surrounding fire. For large fires, use water spray, fog or alcohol-resistant foam. Cool containers with flooding quantities of water until well after fire is out.

Anhydrous ammonia (CAS# 7664-41-7)

COLORLESS GAS OR COLD, MOBILE LIQUID WITH A STRONG, PENETRATING ODOR

NFPA Rating: Health: 3; Flammability: 1; Reactivity: 0

EMERGENCY OVERVIEW
No action shall be taken involving any personal risk or without suitable training. If it is suspected that fumes are still present, the rescuer should wear an appropriate mask or self-contained breathing apparatus. It may be dangerous to the person providing aid to give mouth-to-mouth resuscitation

FIRST AID
EYES: Check for and remove any contact lenses. Immediately flush eyes with plenty of water for at least 15 minutes, occasionally lifting the upper and lower eyelids. Get medical attention immediately.

SKIN: In case of contact, immediately flush skin with plenty of water for at least 15 minutes while removing contaminated clothing and shoes. Wash clothing before reuse. Clean shoes thoroughly before reuse. Get medical attention immediately.
FROSTBITE: Try to warm up the frozen tissues and seek medical attention.
INHALATION: Move exposed person to fresh air. If not breathing, if breathing is irregular or if respiratory arrest occurs, provide artificial respiration or oxygen by trained personnel. Loosen tight clothing such as a collar, tie, belt or waistband. Get medical attention immediately.

Fire hazards
Extremely flammable in the presence of the following materials or conditions: oxidizing materials. Use an extinguishing agent suitable for the surrounding fire.

Protective Equipment and Precautions for Firefighters
As in any fire, wear self-contained breathing apparatus pressure-demand, MSHA/NIOSH (approved or equivalent) and full protective gear.

EXTINGUISHING MEDIA:
Use an extinguishing agent suitable for the surrounding fire

Benzoic acid (CAS#65-85-0)

Physical State: Crystalline powder Appearance: white
Odor: pleasant odor

NFPA Rating: (estimated) Health: 2; Flammability: 1; Reactivity: 0

EMERGENCY OVERVIEW
Causes respiratory tract irritation. Causes severe eye irritation and possible injury. Causes moderate skin irritation. May cause allergic skin reaction. May be harmful if swallowed, inhaled, or absorbed through the skin.

FIRST AID

Eyes: In case of contact, immediately flush eyes with plenty of water for at least 15 minutes. Get medical aid immediately.
Skin: In case of contact, flush skin with plenty of water. Remove contaminated clothing and shoes. Get medical aid if irritation develops and persists. Wash clothing before reuse.
Ingestion: If swallowed, do not induce vomiting unless directed to do so by medical personnel. Never give anything by mouth to an unconscious person. Get medical aid.
Inhalation: If inhaled, remove to fresh air. If not breathing, give artificial respiration. If breathing is difficult, give oxygen. Get medical aid.

Protective Equipment and Precautions for Firefighters
As in any fire, wear self-contained breathing apparatus pressure-demand, MSHA/NIOSH (approved or equivalent) and full protective gear.

Extinguishing Media: Use water spray, dry chemical, carbon dioxide, or chemical foam.

Calcium bromite (CAS#71626-99-8)

Physical State: Solid Appearance: white to off-white Odor: Odorless.

NFPA Rating: (estimated) Health: 2; Flammability: 0; Instability: 0

EMERGENCY OVERVIEW:
Appearance: white to off-white solid.
Causes severe eye irritation and possible eye injury. May cause skin and
respiratory tract irritation.

FIRST AID
Eyes: In case of contact, immediately flush eyes with plenty of water for a t least 15
minutes. Get medical aid immediately. Skin: In case of contact, flush skin with plenty of water. Remove contaminated clothing and shoes. Get medical aid if irritation develops and persists. Wash clothing before reuse.
Ingestion: If swallowed, do not induce vomiting unless directed to do so by medical
personnel. Never give anything by mouth to an unconscious person. Get medical aid.

Inhalation: If inhaled, remove to fresh air. If not breathing, give artificial
respiration. If breathing is difficult, give oxygen. Get medical aid.

Protective Equipment and Precautions for Firefighters
As in any fire, wear self-contained breathing apparatus pressure-demand, MSHA/NIOSH (approved or equivalent) and full protective gear.

Extinguishing Media: Use extinguishing media most appropriate for the
surrounding fire.

Calcium chloride (CAS#10035-04-8)

Physical State: Solid Appearance: white Odor: odorless

NFPA Rating: (estimated) Health: 2; Flammability: 0; Reactivity: 0

EMERGENCY OVERVIEW
May be harmful if swallowed. May cause severe respiratory and digestive tract irritation with possible burns. May cause severe eye and skin irritation with possible burns. May cause cardiac disturbances.

FIRST AID
Eyes: Immediately flush eyes with plenty of water for at least 15 minutes, occasionally lifting the upper and lower eyelids. Get medical aid.
Skin: Get medical aid. Immediately flush skin with plenty of soap and water for at least 15 minutes while removing contaminated clothing and shoes. Wash clothing before reuse.
Ingestion: Do NOT induce vomiting. If victim is conscious and alert, give 2-4 cupfuls of milk or water. Never give anything by mouth to an unconscious person. Get medical aid.
Inhalation: Remove from exposure to fresh air immediately. If not breathing, give artificial respiration. If breathing is difficult, give oxygen. Get medical aid. Do **NOT** use mouth-to-mouth resuscitation.

Protective Equipment and Precautions for Firefighters
As in any fire, wear self-contained breathing apparatus pressure-demand, MSHA/NIOSH (approved or equivalent) and full protective gear.

Extinguishing Media: Use extinguishing media most appropriate for the surrounding fire.

Calcium hydroxide (CAS#1305-62-0)

Appearance Off-white Physical State Solid Odor odorless

NFPA Health 3 Flammability 0 Instability 0

Emergency Overview
Causes skin and eye burns. Risk of serious damage to eyes. Causes respiratory tract irritation and possible burns.

FIRST AID
Eye Contact Rinse immediately with plenty of water, also under the eyelids, for at least 15 minutes. Immediate medical attention is required.
Skin Contact Wash off immediately with plenty of water for at least 15 minutes. Immediate medical attention is required.
Inhalation Move to fresh air. If breathing is difficult, give oxygen. Do not use mouth-to-mouth resuscitation if victim ingested or inhaled the substance; induce artificial respiration with a respiratory medical device. Immediate medical attention is required.
Ingestion Do not induce vomiting. Call a physician or Poison Control Center immediately.

Protective Equipment and Precautions for Firefighters
As in any fire, wear self-contained breathing apparatus pressure-demand, MSHA/NIOSH (approved or equivalent) and full protective gear.

Suitable Extinguishing Media Substance is nonflammable; use agent most appropriate to extinguish surrounding fire..
Unsuitable Extinguishing Media Carbon dioxide (CO2)

Calcium hypochlorite (CAS#17778-54-3)

Appearance: White or grayish-white granules or chips Odor: Chlorine-like

NFPA Health: 3; Flammability; 0; Reactivity: 1; Oxidizer

Emergency Overview

Danger! Strong oxidizer. Contact with other material may cause a fire. Corrosive. Causes eye and skin burns. Harmful if swallowed. Contact with acids liberates toxic gas. May cause severe respiratory tract irritation with possible burns. May cause severe digestive tract irritation with possible burns. Air sensitive.

FIRST AID
Eyes: Get medical aid immediately. Do NOT allow victim to rub eyes or keep eyes closed. Extensive irrigation with water is required (at least 30 minutes).
Skin: Get medical aid immediately. Immediately flush skin with plenty of water for at least 15 minutes while removing contaminated clothing and shoes. Wash clothing before reuse. Destroy contaminated shoes.
Ingestion: Do not induce vomiting. If victim is conscious and alert, give 2-4 cupfuls of milk or water. Never give anything by mouth to an unconscious person. Get medical aid immediately.
Inhalation: Get medical aid immediately. Remove from exposure and move to fresh air immediately. If breathing is difficult, give oxygen. Do NOT use mouth-to-mouth resuscitation. If breathing has ceased apply artificial respiration using oxygen and a suitable mechanical device such as a bag and a mask.

Protective Equipment and Precautions for Firefighters
As in any fire, wear self-contained breathing apparatus pressure-demand, MSHA/NIOSH (approved or equivalent) and full protective gear.

Explosion: Sealed containers may rupture when heated. An explosion can occur if a dry ammonium compound fire extinguisher is used to extinguish a fire involving calcium hypochlorite.
Extinguishing Media: Use water only! Do NOT use carbon dioxide or dry chemical. Contact professional fire-fighters immediately. Cool containers with flooding quantities of water until well after fire is out.

Calcium oxide (CAS#1305-78-8) (QUICKLIME)

Physical State: Solid   Odor & Appearance: Odorless, white powder

NFPA / HMIS Health - 3 Fire - 0 Reactivity - 1 Specific Hazard - ALK

Emergency Overview
This product is not flammable or combustible. This product is not explosive.
May react violently with water or strong acids producing heat and possible steam explosion in confined space.

FIRST AID
Inhalation: Move victim to fresh air. Seek medical attention if necessary. If breathing has stopped, give artificial respiration.
Eyes: Immediately flush eyes with large amounts of water for at least 15 minutes. Pull back the eyelid to make sure all the lime dust has been washed out. Seek medical attention immediately. Do not rub eyes.
Skin: Flush exposed area with large amounts of water. Seek medical attention immediately.
Ingestion: Give large quantities of water or fruit juice. Do not induce vomiting. Seek medical attention immediately. Never give anything by mouth if victim is rapidly losing consciousness or is unconscious or convulsing.

Protective Equipment and Precautions for Firefighters
As in any fire, wear self-contained breathing apparatus pressure-demand, MSHA/NIOSH (approved or equivalent) and full protective gear.
Extinguishing media: Use dry chemical fire extinguisher. Do not use water or halogenated compounds except that large amounts of water may be used to deluge small quantities of lime. Use appropriate extinguishing media for surrounding fire conditions.

Carbon tetrachloride (CAS#56-23-5)

COLOR: colorless PHYSICAL FORM: liquid ODOR: distinct odor

NFPA RATINGS HEALTH=3 FIRE=1 REACTIVITY=0

Emergency Overview
Central nervous system depression.

FIRST AID
INHALATION: If adverse effects occur, remove to uncontaminated area. Give artificial respiration if not breathing. Get immediate medical attention.

SKIN CONTACT: Wash skin with soap and water for at least 15 minutes while removing contaminated clothing and shoes. Get medical attention, if needed. Thoroughly clean and dry contaminated clothing and shoes before reuse.
EYE CONTACT: Flush eyes with plenty of water for at least 15 minutes. Then get immediate medical attention.
INGESTION: If swallowed, drink plenty of water, do NOT induce vomiting. Get immediate medical attention. Induce vomiting only at the instructions of a physician. Do not give anything by mouth to unconscious or convulsive person.

Protective Equipment and Precautions for Firefighters
As in any fire, wear self-contained breathing apparatus pressure-demand, MSHA/NIOSH (approved or equivalent) and full protective gear.

FIRE AND EXPLOSION HAZARDS: Slight fire hazard.
EXTINGUISHING MEDIA: regular dry chemical, regular foam, water

Carbonate dioxide (CAS#124-38-9)

Form : Gas. Color : Colorless. Odor : Odorless. Taste : Acid taste.

NFPA Rating Health: 1 Flammability: 0 Instability: 0

Containers may rupture or explode Potential Health Effects Inhalation : Changes in blood pressure, ringing in the ears, nausea, difficulty breathing, irregular heartbeat, headache, drowsiness, dizziness, tingling sensation, tremors, weakness, visual disturbances, suffocation, convulsions, unconsciousness, coma.

FIRST AID
Eye contact : Contact with liquid: Immediately flush eyes with plenty of water for at least 15 minutes. Then get immediate medical attention.
Skin contact : If frostbite or freezing occur, immediately flush with plenty of lukewarm water (105-115°F; 41-46°C). DO NOT USE HOT WATER. If warm water is not available, gently wrap affected parts in blanket. Get immediate medical attention.
Ingestion : If a large amount is swallowed, get medical attention.
Inhalation : If adverse effects occur, remove to uncontaminated area. Give artificial

respiration if not breathing. If breathing is difficult, oxygen should be administered by qualified personnel. Get immediate medical attention.

Protective Equipment and Precautions for Firefighters
As in any fire, wear self-contained breathing apparatus pressure-demand, MSHA/NIOSH (approved or equivalent) and full protective gear.

Suitable extinguishing media: Use extinguishing agents appropriate for surrounding fire.

Chloroform (CAS#67-6-3)

Appearance: clear, colorless.

NFPA Rating: Health=2, Flammability=0, Reactivity=0

EMERGENCY OVERVIEW
Aspiration hazard. May cause eye and skin irritation. May cause respiratory and digestive tract irritation. May cause cardiac disturbances.

First Aid
Eyes: Immediately flush eyes with plenty of water for at least 15 minutes, occasionally lifting the upper and lower lids. Get medical aid immediately. Do not allow victim to rub or keep eyes closed.
Skin: Get medical aid immediately. Immediately flush skin with plenty of soap and water for at least 15 minutes while removing contaminated clothing and shoes. Discard contaminated clothing in a manner which limits further exposure.
Ingestion: Do not induce vomiting. If victim is conscious and alert, give 2-4 cupfuls of milk or water. Never give anything by mouth to an unconscious person. Possible aspiration hazard. Get medical aid immediately.
Inhalation: Get medical aid immediately. Remove from exposure to fresh air immediately. If breathing is difficult, give oxygen.

Protective Equipment and Precautions for Firefighters
As in any fire, wear self-contained breathing apparatus pressure-demand, MSHA/NIOSH (approved or equivalent) and full protective gear.

Extinguishing Media:
In case of fire, use water fog, dry chemical, carbon dioxide, or regular foam.

Citric acid (CAS#5949-29-1)

Appearance: white crystalline powder. Odorless

NFPA Rating: (estimated) Health: 3; Flammability: 1; Instability: 0

EMERGENCY OVERVIEW
Causes severe eye irritation and possible injury. Causes skin and respiratory tract
irritation.

First Aid
Eyes: In case of contact, immediately flush eyes with plenty of water for a t least 15 minutes. Get medical aid immediately.
Skin: In case of contact, flush skin with plenty of water. Remove contaminated clothing and shoes. Get medical aid if irritation develops and persists. Wash clothing before reuse.
Ingestion: If swallowed, do not induce vomiting unless directed to do so by medical personnel. Never give anything by mouth to an unconscious person. Get medical aid.
Inhalation: If inhaled, remove to fresh air. If not breathing, give artificial respiration. If breathing is difficult, give oxygen. Get medical aid.

Protective Equipment and Precautions for Firefighters
As in any fire, wear self-contained breathing apparatus pressure-demand, MSHA/NIOSH (approved or equivalent) and full protective gear.

Suitable Extinguishing Media Use water spray, dry chemical, carbon dioxide, or chemical foam.

Crude oil (CAS#8002-05-9)
Thick, dark yellow to brown or green-black liquid.
NFPA ratings Health: 2 Flammability: 3 Instability: 0

Emergency overview
WARNING! Flammable. Will be easily ignited by heat, spark or flames.
Aspiration hazard: Harmful or fatal if swallowed. Can enter lungs and cause damage. Causes skin and eye irritation.

First Aid

Inhalation : Move to fresh air. Administer oxygen or artificial respiration if needed. Seek medical attention immediately.
Skin contact : Take off all contaminated clothing immediately. Wash off immediately with soap and plenty of water. Seek medical attention if irritation or skin thermal burns occur.
Eye contact : In case of eye contact, immediately flush with low pressure, cool water for at least 15 minutes, opening eyelids to ensure flushing. Hold the eyelids open and away from the eyeballs to ensure that all surfaces are flushed thoroughly. Seek medical attention immediately.
Ingestion : Do NOT induce vomiting. Never give anything by mouth to an unconscious person. Seek medical attention immediately. If vomiting does occur naturally, keep head below the hips to reduce the risks of aspiration. Small amounts of material which enter the mouth should be rinsed out until the taste is dissipated.

Protective Equipment and Precautions for Firefighters
As in any fire, wear self-contained breathing apparatus pressure-demand, MSHA/NIOSH (approved or equivalent) and full protective gear.

Suitable extinguishing media Water. Water fog. Foam. Dry chemical powder. Carbon dioxide (CO2).

Diesel Fuels (CAS#68476-34-6)

All Grades, Liquid (may be dyed red).

NFPA ratings Health: 2 Flammability: 2 Instability: 0

Emergency overview
Combustible liquid and vapor. May be ignited by heat, sparks or flames. Heat may cause the containers to explode.

First Aid
Eye contact Immediately flush eyes with plenty of water for at least 15 minutes. Remove contact lenses, if present and easy to do. Continue rinsing. Get medical attention.
Skin contact Remove contaminated clothing and shoes. Wash off immediately with soap and plenty of water. Get medical attention if irritation develops or persists. Wash clothing separately before reuse.

Destroy or thoroughly clean contaminated shoes. If high pressure injection under the skin occurs,
always seek medical attention.
Inhalation Move to fresh air. If breathing is difficult, give oxygen. If not breathing, give artificial respiration. Get medical attention.
Ingestion Rinse mouth thoroughly. Do not induce vomiting without advice from poison control center. Do not
give mouth-to-mouth resuscitation. If vomiting occurs, keep head low so that stomach content
does not get into the lungs. Get medical attention immediately.

Protective Equipment and Precautions for Firefighters
As in any fire, wear self-contained breathing apparatus pressure-demand, MSHA/NIOSH (approved or equivalent) and full protective gear.

Fire Fighting Measures
Flammable properties Combustible liquid and vapor. Containers may explode when heated.
Suitable extinguishing media Water spray. Water fog. Foam. Dry chemical powder. Carbon dioxide (CO2).

Ethylene Diamine (CAS#107-15-3)

Physical State: Viscous liquid Color: clear, colorless Odor: ammonia-like

NFPA Rating: health: 3; flammability: 3; instability: 0;

EMERGENCY OVERVIEW
Danger! Flammable liquid and vapor. Harmful if absorbed through the skin
Causes eye and skin burns. Causes digestive and respiratory tract burns. May cause allergic respiratory and skin reaction.

First Aid
Eyes: In case of contact, immediately flush eyes with plenty of water for at least 15 minutes. Get medical aid immediately.
Skin: In case of contact, immediately flush skin with plenty of water for at least 15 minutes while removing contaminated clothing and shoes. Get medical aid immediately. Wash clothing before reuse.
Ingestion: If swallowed, do NOT induce vomiting. Get medical aid immediately. If victim is fully conscious, give a

cupful of water. Never give anything by mouth to an unconscious person.
Inhalation: If inhaled, remove to fresh air. If not breathing, give artificial respiration. If breathing is difficult, give oxygen. Get medical aid.

Protective Equipment and Precautions for Firefighters
As in any fire, wear self-contained breathing apparatus pressure-demand, MSHA/NIOSH (approved or equivalent) and full protective gear.

Extinguishing Media:
Solid streams of water may be ineffective and spread material. Use water spray, dry chemical, "alcohol resistant" foam, or carbon dioxide.
Flammable properties
During a fire, irritating and highly toxic gases may be generated by thermal decomposition or combustion. Use water spray to keep fire-exposed containers cool. Flammable liquid and vapor. Vapors are heavier than air and may travel to a source of ignition and flash back. Vapors can spread along the ground and collect in low or confined areas.

Ethylene glycol (CAS#107-21-1)

Form: liquid Appearance: Turbid Color: Colorless, Pale yellow Odor: slight

NFPA Rating Health 1 Flammability 1 Reactivity 0

EMERGENCY OVERVIEW
May cause eye, skin, and respiratory tract irritation. Use cold water spray to cool fire-exposed containers to minimize the risk of rupture. Harmful if swallowed.

First Aid
Eye contact In case of contact, flush eyes with plenty of lukewarm water.
Skin contact In case of skin contact, wash affected areas with soap and water.
Inhalation If inhaled, remove to fresh air. Get medical attention if irritation develops.
Ingestion If ingested, do not induce vomiting unless directed to do so by medical personnel. Get medical attention.

Protective Equipment and Precautions for Firefighters

As in any fire, wear self-contained breathing apparatus pressure-demand, MSHA/NIOSH (approved or equivalent) and full protective gear.

Suitable extinguishing media: Carbon dioxide (CO_2), Dry chemical, Foam, water spray for large fires.

Ethylene glycol monobutyl ether (CAS#111-76-2)

Colorless liquid

NFPA Rating Health hazard:2 Fire Hazard: 2 Reactivity Hazard: 0

EMERGENCY OVERVIEW
Combustible liquid. Harmful if swallowed, in contact with skin or if inhaled. Causes skin irritation.
Causes serious eye irritation.

First Aid
Inhaled If breathed in, move person into fresh air. If not breathing, give artificial respiration. Consult a physician.
Skin contact Wash off with soap and plenty of water. Consult a physician.
Eye contact Rinse thoroughly with plenty of water for at least 15 minutes and consult a physician.
If swallowed Do NOT induce vomiting. Never give anything by mouth to an unconscious person. Rinse mouth with water. Consult a physician.

Protective Equipment and Precautions for Firefighters
As in any fire, wear self-contained breathing apparatus pressure-demand, MSHA/NIOSH (approved or equivalent) and full protective gear.

Suitable extinguishing media
Use water spray, alcohol-resistant foam, dry chemical or carbon dioxide.

Ethylenediaminetetraacetic acid (EDTA) (CAS#60-00-4)

Colorless to white odorless crystals

NFPA Rating: health: 2; flammability: 1; instability: 0;

First Aid

Eyes: Immediately flush eyes with plenty of water for at least 15 minutes, occasionally lifting the upper and lower eyelids. Get medical aid.
Skin: Get medical aid. Immediately flush skin with plenty of water for at least 15 minutes while removing contaminated clothing and shoes. Wash clothing before reuse.
Ingestion: Never give anything by mouth to an unconscious person. Get medical aid. Do NOT induce vomiting. If conscious and alert, rinse mouth and drink 2-4 cupfuls of milk or water.
Inhalation: Remove from exposure and move to fresh air immediately. If not breathing, give artificial respiration. If breathing is difficult, give oxygen. Get medical aid.

Protective Equipment and Precautions for Firefighters
As in any fire, wear self-contained breathing apparatus pressure-demand, MSHA/NIOSH (approved or equivalent) and full protective gear.

Extinguishing Media: Use water spray, dry chemical, carbon dioxide, or appropriate foam.

Formic acid (CAS#64-18-6)

Appearance: Colorless Odor: Pungent

NFPA Rating: Health 3 Flammability 2 Instability 0

EMERGENCY OVERVIEW
Flammable liquid and vapor. Strong reducing agent. Fire and explosion risk in contact with oxidizing agents. Causes severe burns by all exposure routes. Hygroscopic.

First Aid
Eyes: Get medical aid immediately. Immediately rinse eyes with plenty of water for at least 15 minutes.
Skin: Wash off immediately with plenty of water for at least 15 minutes. Get medical aid immediately.
Ingestion: Do NOT induce vomiting. Dilute with 2-4 cupfuls of milk or water. Get medical aid immediately.
Inhalation: Get medical aid immediately. Remove from exposure to fresh air immediately. If breathing is difficult, give oxygen. Do NOT use mouth-to-mouth resuscitation. If breathing has ceased apply artificial respiration using oxygen and a suitable mechanical device such as a bag and a mask.

Protective Equipment and Precautions for Firefighters
As in any fire, wear self-contained breathing apparatus pressure-demand, MSHA/NIOSH (approved or equivalent) and full protective gear.

Suitable Extinguishing Media CO2, dry chemical, dry sand, alcohol-resistant foam. Cool closed containers exposed to fire with water spray.

Gasoline (CAS#8006-61-9)

PHYSICAL STATE: Clear, Bright Liquid/Vapor ODOR: Petroleum hydrocarbon, Gasoline odor

NFPA Rating: Health:1 ; Flammability:3; Reactivity ;0

EMERGENCY OVERVIEW
Danger! This product is a highly flammable liquid and vapor. This product contains petroleum hydrocarbons that can pose health risks. Avoid liquid, mist and vapor contact. Harmful or fatal if swallowed. Aspiration hazard can enter lungs and cause damage. May cause irritation or be harmful if inhaled or absorbed through the skin. Avoid prolonged or repeated skin contact.

FIRST AID
EYES: Flush immediately with large amounts of water for at least 15 minutes. Eyelids should be held away from the eyeball to ensure thorough rinsing. Seek medical advice if pain or redness continues.
SKIN: In case of contact, immediately flush skin with large amounts of water for at least 15 minutes. Remove contaminated clothing promptly and launder before reuse. Contaminated leather goods should be discarded. Wash exposed area thoroughly with soap and water. If irritation persists or symptoms described in the MSDS develop, seek medical attention.
INGESTION: This material may cause nausea, vomiting, diarrhea and restlessness. DO NOT INDUCE VOMITING. Aspiration into the lungs presents a significant chemical pneumonitis hazard. Obtain medical attention promptly. If vomiting occurs spontaneously, keep head below hips to prevent aspiration.
INHALATION: If inhaled, remove to fresh air. If breathing is difficult, give oxygen.

Protective Equipment and Precautions for Firefighters

As in any fire, wear self-contained breathing apparatus pressure-demand, MSHA/NIOSH (approved or equivalent) and full protective gear.

FIRE-FIGHTING MEDIA AND INSTRUCTIONS:
Flammable liquid and vapor. Use dry chemical, foam, or carbon dioxide to extinguish the fire. Consult foam manufacturer for appropriate media, application rates and water/foam ratio. Water spray may be used to disperse vapors and/or flush spills away from source of ignition. Water spray can be used to cool tanks and exposures. Vapor suppressing foam may be used to suppress vapors. Vapor may cause flash fire. Vapors are heavier than air and may accumulate in low or confined areas or travel a considerable distance to a source of ignition and flashback. Runoff to sewer may create fire or explosion hazard.

Gilsonite (CAS#12002-43-6)

APPEARANCE/PHYSICAL STATE: Powder, dust. COLOR: Black. ODOR: Hydrocarbon.

NFPA: Unavailable

EMERGENCY OVERVIEW:
CAUTION! MAY CAUSE EYE, SKIN AND RESPIRATORY TRACT IRRITATION.

FIRST AID
INHALATION: Move the exposed person to fresh air at once. Perform artificial respiration if breathing has stopped. Get medical attention.
INGESTION: Drink a couple of glasses water or milk. Do NOT induce vomiting unless directed to do so by a physician. Never give anything by mouth to an unconscious person. Get medical attention.
SKIN: Wash skin thoroughly with soap and water. Remove contaminated clothing. Get medical attention if any discomfort continues.
EYES: Promptly wash eyes with lots of water while lifting the eye lids. Continue to rinse for at least 15 minutes. Get medical attention if any discomfort continues.

Protective Equipment and Precautions for Firefighters

As in any fire, wear self-contained breathing apparatus pressure-demand, MSHA/NIOSH (approved or equivalent) and full protective gear.

Firefighting
Fires involving this material can be controlled with a dry chemical, carbon dioxide or Halon extinguisher. A water fog may also be used

Glutaraldehyde (CAS#111-30-8)

Appearance: clear, colorless liquid.

NFPA Rating: (estimated) Health: 3; Flammability: 1; Instability: 1

EMERGENCY OVERVIEW
Danger! Causes eye and skin burns. Causes digestive and respiratory tract burns. May cause allergic respiratory and skin reaction. Harmful if swallowed, inhaled, or absorbed through the skin. Aspiration hazard if swallowed. Can enter lungs and cause damage.

FIRST AID
Eyes: Immediately flush eyes with plenty of water for at least 15 minutes, occasionally lifting the upper and lower eyelids. Get medical aid immediately.
Skin: Get medical aid immediately. Immediately flush skin with plenty of water for at least 15 minutes while removing contaminated clothing and shoes. Wash clothing before reuse. Destroy contaminated shoes.
Ingestion: Do not induce vomiting. Get medical aid immediately. Call a poison control center. Do not give anything by mouth. Oral toxicity of Glutaraldehyde increases with dilution. Drinking water following
ingestion of concentrated Glutaraldehyde solutions can enhance the toxicity.
Inhalation: Get medical aid immediately. Remove from exposure and move to fresh air immediately. If breathing is difficult, give oxygen. Do NOT use mouth-to-mouth resuscitation. If breathing has ceased apply artificial respiration using oxygen and a suitable mechanical device such as a bag and a mask.

Protective Equipment and Precautions for Firefighters
As in any fire, wear self-contained breathing apparatus pressure-demand, MSHA/NIOSH (approved or equivalent) and full protective gear.

Extinguishing Media: Use water spray, dry chemical, carbon dioxide, or appropriate foam.

Hydrochloric acid (CAS#7647-01-0)

Appearance Colorless Physical State Liquid Odor pungent

NFPA Rating: Health 3 Flammability 0 Instability 1

Emergency Overview
Corrosive to metals. Causes burns by all exposure routes. May cause irritation of respiratory tract.

First Aid
Eye Contact Rinse immediately with plenty of water, also under the eyelids, for at least 15 minutes. Immediate medical attention is required.
Skin Contact Wash off immediately with plenty of water for at least 15 minutes. Immediate medical attention is required.
Inhalation Move to fresh air. If breathing is difficult, give oxygen. Do not use mouth-to-mouth resuscitation if victim ingested or inhaled the substance; induce artificial respiration with a respiratory medical device. Immediate medical attention is required.
Ingestion Do not induce vomiting. Call a physician or Poison Control Center immediately.

Protective Equipment and Precautions for Firefighters
As in any fire, wear self-contained breathing apparatus pressure-demand, MSHA/NIOSH (approved or equivalent) and full protective gear.

Suitable Extinguishing Media Substance is nonflammable; use agent most appropriate to extinguish surrounding fire..

Hydrofluoric acid (CAS#7664-39-3)

Appearance: colorless Physical State: Clear liquid Odor: strong odor

NFPA Rating: Health 4 Flammability 0 Instability 1

EMERGENCY OVERVIEW

Corrosive. Causes severe skin and eye burns. Causes digestive tract burns. May be fatal if inhaled, absorbed through skin, or swallowed. Mist or vapor extremely irritating to eyes and respiratory tract. Causes blood, cardiovascular system and respiratory system damage. Prolonged exposure may cause chronic effects. Reacts with water.

First Aid
Eyes: Get medical aid immediately. Gently lift eyelids and flush continuously with water. Eye exposure may be treated by irrigation with 1% calcium gluconate drops after immediate and copious irrigation with water for at least 30 minutes.
Skin: Get medical aid immediately. Rinse area with large amounts of water for at least 15 minutes. Remove contaminated clothing and shoes. For exposures to hydrofluoric acid concentrations less than 20%, liberal and frequent applications of a 2.5% calcium gluoconate gel may be applied.
Ingestion: Do NOT induce vomiting. If victim is conscious and alert, give 2-4 cupfuls of milk or water. Get medical aid immediately.
Inhalation: Remove from exposure and move to fresh air immediately. If not breathing, give artificial respiration. If breathing is difficult, give oxygen. Get medical aid.

Protective Equipment and Precautions for Firefighters
As in any fire, wear self-contained breathing apparatus pressure-demand, MSHA/NIOSH (approved or equivalent) and full protective gear.

Suitable Extinguishing Media Use an extinguishing agent suitable for the surrounding fire. Apply water from a safe distance to cool container and protect surrounding area. If involved in fire, shut off flow immediately if it can be done without risk. Contains gas under pressure. In a fire or if heated, a pressure increase will occur and the container may burst or explode.

Hydrogen Sulfide "H2S" (CAS#7783-06-04)

(See H2S Section for more detailed information)

PHYSICAL STATE: gas COLOR: colorless ODOR: rotten egg odor TASTE: sweet taste (Odor is not a reliable indicator of hydrogen sulfide's presence and may not provide adequate warning of hazardous concentrations) . Although very pungent at first, it quickly deadens the sense of smell, so potential victims may be unaware of its presence until it is too late.

NFPA RATINGS: HEALTH=4 FIRE=4 REACTIVITY=0

EMERGENCY OVERVIEW:
Hydrogen sulfide is a **highly** toxic and **flammable** gas. Being heavier than air, it tends to accumulate at the bottom of poorly ventilated spaces. Short-term, high-level exposure can induce immediate collapse, with a high probability of death.
Hot Zone Rescuers should be trained and appropriately attired before entering the Hot Zone. Rescuers should have a **safety line** during rescue operations because of the extremely rapid toxic action of hydrogen sulfide.

INHALATION:
SHORT TERM EXPOSURE: irritation, cough, lack of sense of smell, sensitivity to light, changes in blood pressure, nausea, vomiting, difficulty breathing, headache, drowsiness, dizziness, disorientation, hallucinations, pain in extremities, tremors, visual disturbances, suffocation, lung congestion, internal bleeding, heart disorders, nerve damage, brain damage, convulsions, coma, death
LONG TERM EXPOSURE: loss of appetite, weight loss, irregular heartbeat, headache, sleep disturbances, lung congestion, nerve damage, paralysis, effects on the brain

FIRST AID MEASURES
INHALATION: If adverse effects occur, remove to uncontaminated area. Give artificial respiration if not breathing. If breathing is difficult, oxygen should be administered by qualified personnel. Get immediate medical attention.
SKIN CONTACT: Wash skin with soap and water for at least 15 minutes while removing contaminated clothing and shoes. Get medical attention, if needed. Thoroughly clean and dry contaminated clothing and shoes before reuse.

EYE CONTACT: Flush eyes with plenty of water for at least 15 minutes. Then get immediate medical attention.

INGESTION: : ingestion of a gas is unlikely

Protective Equipment and Precautions for Firefighters
As in any fire, wear self-contained breathing apparatus pressure-demand, MSHA/NIOSH (approved or equivalent) and full protective gear.

FIRE AND EXPLOSION HAZARDS: Severe fire hazard. The vapor is heavier than air. Vapors or gases may ignite at distant ignition sources and flash back. Pressurized containers may rupture or explode if exposed to sufficient heat. Electrostatic discharges may be generated by flow or agitation resulting in ignition or explosion.

EXTINGUISHING MEDIA: Let burn unless leak can be stopped immediately. Large fires: Use regular foam or flood with fine water spray.

FIRE FIGHTING: Move container from fire area if it can be done without risk. Withdraw immediately in case of rising sound from venting safety device or any discoloration of tanks due to fire. Cool containers with water spray until well after the fire is out. Keep unnecessary people away, isolate hazard area and deny entry.
For tank, rail car or tank truck, evacuation radius: Evacuation radius: 800 meters (1/2 mile). Do not attempt to extinguish fire unless flow of material can be stopped first. Flood with fine water spray. Do not scatter spilled material with high-pressure water streams. Cool containers with water. Apply water from a protected location or from a safe distance. Avoid inhalation of material or combustion by-products. Stay upwind and keep out of low areas. Stop flow of gas.

Isopropanol (CAS#67-63-0)

Appearance: colorless liquid.

NFPA Rating: (estimated) Health: 1; Flammability: 3; Reactivity: 0

EMERGENCY OVERVIEW
May cause central nervous system depression. May form explosive peroxides. **Flammable liquid and vapor.** Hygroscopic. Causes respiratory tract irritation. Aspiration hazard if swallowed. Can enter lungs and cause damage. This material has been reported to be susceptible to autoxidation and therefore should be classified as peroxidizable. Causes eye irritation. Breathing vapors may cause drowsiness and dizziness. Prolonged or repeated contact causes defatting of the skin with irritation, dryness, and cracking.

FIRST AID

Eyes: In case of contact, immediately flush eyes with plenty of water for at least 15 minutes. Get medical aid.
Skin: In case of contact, flush skin with plenty of water. Remove contaminated clothing and shoes. Get medical aid if irritation develops and persists. Wash clothing before reuse.
Ingestion: Potential for aspiration if swallowed. Get medical aid immediately. Do not induce vomiting unless directed to do so by medical personnel. Never give anything by mouth to an unconscious person.
Inhalation: If inhaled, remove to fresh air. If not breathing, give artificial respiration. If breathing is difficult, give oxygen. Get medical aid.

Protective Equipment and Precautions for Firefighters
As in any fire, wear self-contained breathing apparatus pressure-demand, MSHA/NIOSH (approved or equivalent) and full protective gear.

Extinguishing Media: Water may be ineffective. Do NOT use straight streams of water. For large fires, use dry chemical, carbon dioxide, alcohol-resistant foam, or water spray. For small fires, use carbon dioxide, dry chemical, dry sand, or alcohol-resistant foam. Cool containers with flooding quantities of water until well after fire is out.

Kerosene (CAS#8008-20-6)

Appearance : liquid Color : yellow Odor : petroleum hydrocarbon odor

NFPA Rating: (estimated) Health: 2; Flammability: 2; Reactivity: 0

EMERGENCY OVERVIEW
Flammable. Irritating to skin.Harmful: may cause lung damage if swallowed.

FIRST AID
Inhalation : Keep at rest. Move to fresh air. Oxygen or artificial respiration if needed. Call a physician immediately.
Skin contact : Wash off immediately with soap and plenty of water.
Take off contaminated clothing and shoes immediately. Wash contaminated clothing before re-use. If skin irritation persists, call a physician.

Eye contact : Rinse immediately with plenty of water, also under the eyelids, for at least 15 minutes. If eye irritation persists, consult a specialist.
Ingestion : Do NOT induce vomiting. Drink plenty of water. Never give anything by mouth to an unconscious person. Consult a physician.

Protective Equipment and Precautions for Firefighters
As in any fire, wear self-contained breathing apparatus pressure-demand, MSHA/NIOSH (approved or equivalent) and full protective gear.

Suitable extinguishing media : Use dry chemical, CO2, water spray or alcohol resistant foam

Lime (CAS#1305-78-8)

Appearance Granular. Color White to yellow.

NFPA Health: 3 Flammability: 0 Instability: 1 Special hazards: W

Emergency overview DANGER
Corrosive. Causes severe skin and eye burns. Causes digestive tract burns. Mist or vapor
extremely irritating to eyes and respiratory tract. Reacts with water.

FIRST AID
Eye contact Immediately flush with plenty of water for at least 15 minutes. If easy to do, remove contact lenses. Call a physician or poison control center immediately. In case of irritation from airborne exposure, move to fresh air. Get medical attention immediately.
Skin contact Immediately flush with plenty of water for at least 15 minutes while removing contaminated clothing and shoes. Call a physician or poison control center immediately. Wash clothing separately before reuse. Destroy or thoroughly clean contaminated shoes.
Inhalation Move to fresh air. If breathing stops, provide artificial respiration. If breathing is difficult, give oxygen. Call a physician or poison control center immediately.
Ingestion Call a physician or poison control center immediately. Do not induce vomiting. If vomiting occurs, the head should be kept low so that stomach vomit doesn't enter the lungs.

Protective Equipment and Precautions for Firefighters

As in any fire, wear self-contained breathing apparatus pressure-demand, MSHA/NIOSH (approved or equivalent) and full protective gear.

Extinguishing Method
Use dry chemical fire extinguisher. Do not use water except in those cases that water may be used to deluge small amounts of Burnt Lime.

Special Fire Fighting Procedures Reaction with water may produce enough heat to ignite combustible materials.

Methanol (CAS#67-56-1)

Clear liquid, slight alcoholic odor

NFPA : (estimated) Health: 1; Flammability: 3; Instability: 0

EMERGENCY OVERVIEW: This product is a volatile, flammable liquid. Highly flammable. Vapors may form explosive mixtures with air.

First Aid
SKIN: Wash off immediately with soap and plenty of water. Take off contaminated clothing and shoes immediately. Wash contaminated clothing before re-use. Call a physician immediately.
EYES: Rinse immediately with plenty of water, also under the eyelids, for at least 15 minutes. Call a physician immediately.
INHALATION: Move to fresh air in case of accidental inhalation of vapours. If not breathing, give artificial respiration. If breathing is difficult, give oxygen, provided a qualified operator is available. Call a physician immediately.
INGESTION: DO NOT induce vomiting. Immediate medical attention is required.

Protective Equipment and Precautions for Firefighters
As in any fire, wear self-contained breathing apparatus pressure-demand, MSHA/NIOSH (approved or equivalent) and full protective gear.

EXTINGUISHING MEDIA:
Use alcohol-resistant foam, carbon dioxide (CO2) or dry chemical.
UNUSUAL FIRE AND EXPLOSION HAZARDS:

Highly flammable. Vapours may form explosive mixtures with air. Vapours are heavier than air and may travel along the ground to some distant source of ignition and flash back. Suppress (knock down) gases/vapours/mists with a water spray jet.

Methyl ethyl ketone (MEK) (CAS#78-93-3)

Physical State: Liquid **Appearance:** colorless **Odor:** sweetish odor - alcohol-like

NFPA Rating: (estimated) Health: 1; Flammability: 3; Instability: 0

EMERGENCY OVERVIEW
Extremely flammable liquid and vapor. Vapor may cause flash fire. May cause respiratory tract irritation. May cause severe eye and skin irritation with possible burns. May cause central nervous system effects.

First Aid
Eyes: Flush eyes with plenty of water for at least 15 minutes, occasionally lifting the upper and lower eyelids. Get medical aid immediately. Do NOT allow victim to rub eyes or keep eyes closed.
Skin: Get medical aid. Rinse area with large amounts of water for at least 15 minutes. Remove contaminated clothing and shoes.
Ingestion: Do not induce vomiting. If victim is conscious and alert, give 2-4 cupfuls of milk or water. Get medical aid immediately.
Inhalation: Get medical aid immediately. Remove from exposure and move to fresh air immediately. If not breathing, give artificial respiration. If breathing is difficult, give oxygen. Do NOT use mouth-to-mouth resuscitation. If breathing has ceased apply artificial respiration using oxygen and a suitable mechanical device such as a bag and a mask.

Protective Equipment and Precautions for Firefighters
As in any fire, wear self-contained breathing apparatus pressure-demand, MSHA/NIOSH (approved or equivalent) and full protective gear.

Extinguishing Media: For small fires, use dry chemical, carbon dioxide, water spray or alcohol-resistant foam. For large fires, use water spray, fog, or alcohol-resistant foam. Do NOT use straight streams of water. Cool containers with flooding quantities of water until well after fire is out. **Flash Point:** -7 deg C (19.40 deg F)

Methyl isobutyl ketone (MIBK) (CAS#108-10-1)

Physical State: Liquid Appearance: clear, colorless Odor: Sweet, camphor-like.

NFPA Rating: (estimated) Health: 2; Flammability: 3; Instability: 0

EMERGENCY OVERVIEW
Flammable liquid and vapor. May cause liver damage. May cause central nervous system depression.

First Aid
Eyes: In case of contact, immediately flush eyes with plenty of water for at least 15 minutes. Get medical aid.
Skin: In case of contact, flush skin with plenty of water. Remove contaminated clothing and shoes. Get medical aid if irritation develops and persists. Wash clothing before reuse.
Ingestion: Potential for aspiration if swallowed. Get medical aid immediately. Do not induce vomiting unless directed to do so by medical personnel. Never give anything by mouth to an unconscious person.
Inhalation: If inhaled, remove to fresh air. If not breathing, give artificial respiration. If breathing is difficult, give oxygen. Get medical aid.

Protective Equipment and Precautions for Firefighters
As in any fire, wear self-contained breathing apparatus pressure-demand, MSHA/NIOSH (approved or equivalent) and full protective gear.

Extinguishing Media: Water may be ineffective. In case of fire, use carbon dioxide, dry chemical powder or appropriate foam

Methylene chloride (CAS#75-09-2)

Physical State: Liquid Appearance: Colorless liquid. Odor: ethereal odor

NFPA Rating: health-2; flammability-1; reactivity-0

EMERGENCY OVERVIEW
May cause respiratory tract irritation. May cause digestive tract irritation. May be harmful if swallowed. May cause central nervous system depression.

First Aid
Eyes: Immediately flush eyes with plenty of water for at least 15 minutes, occasionally lifting the upper and lower lids. Get medical aid immediately.
Skin: Get medical aid. Immediately flush skin with plenty of soap and water for at least 15 minutes while removing contaminated clothing and shoes.
Ingestion: If victim is conscious and alert, give 2-4 cupfuls of milk or water. Never give anything by mouth to an unconscious person. Get medical aid immediately.
Inhalation: Get medical aid immediately. Remove from exposure to fresh air immediately. If not breathing, give artificial respiration. If breathing is difficult, give oxygen.

Protective Equipment and Precautions for Firefighters
As in any fire, wear self-contained breathing apparatus pressure-demand, MSHA/NIOSH (approved or equivalent) and full protective gear.

Extinguishing Media: In case of fire, use water, dry chemical, chemical foam, or alcohol-resistant foam. Use water spray to cool fire-exposed containers.

Naphtha (CAS#8032-32-4)

Clear, colorless liquid with pleasant gasoline odor

NFPA Health 1 Flammability 3 Instability 0

Emergency Overview
Flammable liquid and vapor. May cause eye, skin, and respiratory tract irritation . Inhalation may cause central nervous system effects. Aspiration hazard if swallowed - can enter lungs and cause damage.

First aid
Eye Contact Rinse immediately with plenty of water, also under the eyelids, for at least 15 minutes. Obtain medical attention.

Skin Contact Wash off immediately with plenty of water for at least 15 minutes. Get medical attention immediately if symptoms occur.
Inhalation Move to fresh air. If breathing is difficult, give oxygen. Do not use mouth-to-mouth resuscitation if victim ingested or inhaled the substance; induce artificial respiration with a respiratory medical device. Get medical attention immediately if symptoms occur.
Ingestion Do not induce vomiting. Call a physician or Poison Control Center immediately.

Protective Equipment and Precautions for Firefighters
As in any fire, wear self-contained breathing apparatus pressure-demand, MSHA/NIOSH (approved or equivalent) and full protective gear.

Suitable Extinguishing Media CO2, dry chemical, dry sand, alcohol-resistant foam. Cool closed
containers exposed to fire with water spray.

Naphthalene (CAS#91-20-3)

Physical State: Crystalline powder Appearance: white Odor: Distinctive mothball-like.

NFPA Rating: (estimated) Health: 2; Flammability: 2; Reactivity: 0

EMERGENCY OVERVIEW
May cause allergic skin reaction. Harmful if swallowed. May be fatal if inhaled. Causes eye and skin irritation. Causes digestive and respiratory tract irritation.

First Aid

Eyes: Immediately flush eyes with plenty of water for at least 15 minutes, occasionally lifting the upper and lower eyelids. Get medical aid.
Skin: Get medical aid. Immediately flush skin with plenty of soap and water for at least 15 minutes while removing contaminated clothing and shoes. Wash clothing before reuse.
Ingestion: If swallowed, do not induce vomiting unless directed to do so by medical personnel. Never give anything by mouth to an unconscious person. Get medical aid.
Inhalation: Remove from exposure to fresh air immediately. If breathing is difficult, give oxygen. Get medical aid. Do NOT use mouth-to-mouth resuscitation. If breathing has ceased apply artificial respiration using oxygen and a suitable mechanical device such as a bag and a mask.

Protective Equipment and Precautions for Firefighters
As in any fire, wear self-contained breathing apparatus pressure-demand, MSHA/NIOSH (approved or equivalent) and full protective gear.

Extinguishing Media: Use dry sand or earth to smother fire. Water or foam may cause frothing. Cool containers with flooding quantities of water until well after fire is out. Use dry chemical, carbon dioxide, or appropriate foam.

Nitrogen (CAS#7727-37-9)

Gas. [NORMALLY A COLORLESS GAS: MAY BE A CLEAR COLORLESS LIQUID AT LOW TEMPERATURES. SOLD AS A COMPRESSED GAS OR LIQUID IN STEEL CYLINDERS.]

NFPA: **Gas** Health: 0 Flammability:0 Physical Hazards: 0
Liquid Health: 3 Flammability:0 Physical Hazards: 0

Emergency overview
GAS:CONTENTS UNDER PRESURE. Can cause rapid suffocation. May cause severe frostbite.
LIQUID: Extremely cold liquid and gas under pressure. Can cause rapid suffocation. May cause severe frostbite. It may be dangerous to the person providing aid to give mouth-to-mouth resuscitation.

First aid
Eye contact: Check for and remove any contact lenses. Immediately flush eyes with plenty of water for at least 15 minutes, occasionally lifting the upper and lower eyelids. Get medical attention immediately.

Frostbite : Try to warm up the frozen tissues and seek medical attention.
Inhalation: Move exposed person to fresh air. If not breathing, if breathing is irregular or if respiratory arrest occurs, provide artificial respiration or oxygen by trained personnel. Loosen tight clothing such as a collar, tie, belt or waistband. Get medical attention immediately.

Protective Equipment and Precautions for Firefighters
As in any fire, wear self-contained breathing apparatus pressure-demand, MSHA/NIOSH (approved or equivalent) and full protective gear.

Fire-fighting media: Use an extinguishing agent suitable for the surrounding fire.

Oxalic acid (CAS#144-62-7)

Physical State: Powder Appearance: white Odor: odorless

NFPA Rating: (estimated) Health: 3; Flammability: 1; Reactivity: 0

Emergency overview
May cause severe respiratory tract irritation with possible burns. May cause severe digestive tract irritation with possible burns. May cause kidney damage. May cause eye and skin irritation with possible burns. Harmful in contact with skin and if swallowed.

First aid
Eyes: In case of contact, immediately flush eyes with plenty of water for at least 15 minutes. Get medical aid immediately.
Skin: In case of contact, immediately flush skin with plenty of water for at least 15 minutes while removing contaminated clothing and shoes. Get medical aid immediately. Wash clothing before reuse.
Ingestion: If swallowed, do NOT induce vomiting. Get medical aid immediately. If victim is fully conscious, give a cupful of water. Never give anything by mouth to an unconscious person.
Inhalation: If inhaled, remove to fresh air. If not breathing, give artificial respiration. If breathing is difficult, give oxygen. Get medical aid.

Protective Equipment and Precautions for Firefighters

As in any fire, wear self-contained breathing apparatus pressure-demand, MSHA/NIOSH (approved or equivalent) and full protective gear.

General Information: As in any fire, wear a self-contained breathing apparatus in pressure-demand, MSHA/NIOSH (approved or equivalent), and full protective gear. During a fire, irritating and highly toxic gases may be generated by thermal decomposition or combustion. Use water spray to keep fire-exposed containers cool.
Extinguishing Media: Use water spray, dry chemical, carbon dioxide, or alcohol-resistant foam.

Paraformaldehyde (CAS#30525-89-4)

Form (physical state): Solid Color: White crystalline

NFPA Health 2 Flammability 3 Instability 0

Emergency Overview
Flammable solid. Harmful by inhalation and if swallowed. Irritating to eyes, respiratory system and skin

FIRST AID
Eye Contact Rinse immediately with plenty of water, also under the eyelids, for at least 15 minutes. Immediate medical attention is required.
Skin Contact Wash off immediately with plenty of water for at least 15 minutes. Obtain medical attention.
Inhalation Move to fresh air. If breathing is difficult, give oxygen. Do not use mouth-to-mouth resuscitation if victim ingested or inhaled the substance; induce artificial

Protective Equipment and Precautions for Firefighters
As in any fire, wear self-contained breathing apparatus pressure-demand, MSHA/NIOSH (approved or equivalent) and full protective gear.

Suitable Extinguishing Media Use water spray, alcohol-resistant foam, dry chemical or carbon dioxide. Cool closed containers exposed to fire with water spray.

Petroleum distillate (CAS#8030-30-6)

Form : Liquid Appearance : Colorless to light yellow Odor : Characteristic hydrocarbon-like

NFPA Health 1 Flammability 3 Instability 0

Hazard Summary :
Extremely flammable. Irritating to eyes and respiratory system. Affects central
nervous system. Harmful or fatal if swallowed. Aspiration Hazard.

FIRST AID
General advice : Remove from exposure, lie down. In the case of accident or if you feel unwell, seek medical advice immediately (show the label where possible). When symptoms persist or in all cases of doubt, seek medical advice. Never give
anything by mouth to an unconscious person. Take off all contaminated clothing
immediately and thoroughly wash material from skin.
Inhalation : If inhaled, remove to fresh air. If not breathing, give artificial respiration. If breathing is difficult, give oxygen. Seek medical attention immediately.
Skin contact : In case of contact, immediately flush skin with plenty of water. Take off contaminated clothing and shoes immediately. Wash contaminated clothing
before re-use. Contaminated leather, particularly footwear, must be discarded.
Note that contaminated clothing may be a fire hazard. Seek medical advice if
symptoms persist or develop.
Eye contact : Remove contact lenses. In the case of contact with eyes, rinse immediately with plenty of water and seek medical advice.
Ingestion : If swallowed Do NOT induce vomiting. Never give anything by mouth to an unconscious person. Seek medical attention immediately.

Protective Equipment and Precautions for Firefighters
As in any fire, wear self-contained breathing apparatus pressure-demand, MSHA/NIOSH (approved or equivalent) and full protective gear.

Suitable extinguishing media : Use water spray, alcohol-resistant foam, dry chemical or carbon dioxide. Do not
use a solid water stream as it may scatter and spread fire.

Propane (CAS#74-98-6)

Gas. [COLORLESS LIQUEFIED COMPRESSED GAS; ODORLESS BUT MAY HAVE

SKUNK ODOR ADDED.]

NFPA Rating: (estimated) Health: 2; Flammability: 4; Reactivity: 0

EMERGENCY OVERVIEW
FLAMMABLE GAS. MAY CAUSE FLASH FIRE. CONTENTS UNDER PRESSURE.

FIRST AID
EYES: Check for and remove any contact lenses. Immediately flush eyes with plenty of water for at least 15 minutes, occasionally lifting the upper and lower eyelids. Get medical attention immediately.
SKIN: In case of contact, immediately flush skin with plenty of water for at least 15 minutes while removing contaminated clothing and shoes. To avoid the risk of static discharges and gas ignition, soak contaminated clothing thoroughly with water before removing it. Wash clothing before reuse. Clean shoes thoroughly before reuse. Get medical attention immediately.
INHALATION: Move exposed person to fresh air. If not breathing, if breathing is irregular or if respiratory arrest occurs, provide artificial respiration or oxygen by trained personnel. Loosen tight clothing such as a collar, tie, belt or waistband. Get medical attention Immediately
FROSTBITE: Try to warm up the frozen tissues and seek medical attention.

Protective Equipment and Precautions for Firefighters
As in any fire, wear self-contained breathing apparatus pressure-demand, MSHA/NIOSH (approved or equivalent) and full protective gear.

Suitable extinguishing media: In case of fire, use water spray (fog), foam or dry chemical. Allow gas to burn if flow cannot be shut off immediately. Apply water from a safe distance to cool container and protect surrounding area. If involved in fire, shut off flow immediately if it can be done without risk. Contains gas under pressure. **Flammable gas**. In a fire or if heated, a pressure increase will occur and the container may burst, with the risk of a subsequent explosion

Potassium hydroxide (CAS#1310-58-3)

Physical State: Solid Appearance: white or yellow
Odor: odorless

NFPA Rating: (estimated) Health: 3; Flammability: 0; Reactivity: 1

EMERGENCY OVERVIEW
Corrosive. Water-Reactive. Harmful if swallowed. Causes severe eye and skin burns. Causes severe digestive and respiratory tract burns.

FIRST AID
Eyes: Immediately flush eyes with plenty of water for at least 15 minutes, occasionally lifting the upper and lower eyelids. Get medical aid immediately.
Skin: Get medical aid immediately. Immediately flush skin with plenty of soap and water for at least 15 minutes while removing contaminated clothing and shoes. Discard contaminated clothing in a manner which limits further exposure.
Ingestion: Do NOT induce vomiting. If victim is conscious and alert, give 2-4 cupfuls of milk or water. Never give anything by mouth to an unconscious person. Get medical aid immediately.
Inhalation: Get medical aid immediately. Remove from exposure to fresh air immediately. If breathing is difficult, give oxygen. If breathing has ceased apply artificial respiration using oxygen and a suitable mechanical device such as a bag and a mask.

Protective Equipment and Precautions for Firefighters
As in any fire, wear self-contained breathing apparatus pressure-demand, MSHA/NIOSH (approved or equivalent) and full protective gear.

Extinguishing Media: For small fires, use dry chemical, carbon dioxide, water spray or alcohol-resistant foam.

Soda ash (CAS#497-19-8)

Physical State: Solid   Appearance: white   Odor: odorless

NFPA/HMIS Health Hazard 2 Flammability 0 Stability 0

EMERGENCY OVERVIEW
Harmful if inhaled. May cause eye and skin irritation with possible burns. May cause respiratory and digestive tract irritation.

FIRST AID
Soda ash is not classified as toxic, but can injure the eyes and irritate the skin
Eyes: Immediately flush eyes with plenty of water for at least 15 minutes, occasionally lifting the upper and lower eyelids. Get medical aid immediately.
Skin: Get medical aid. Flush skin with plenty of soap and water for at least 15 minutes while removing contaminated clothing and shoes. Wash clothing before reuse.
Ingestion: Do NOT induce vomiting. If victim is conscious and alert, give 2-4 cupfuls of milk or water. Never give anything by mouth to an unconscious person. Get medical aid immediately.
Inhalation: Remove from exposure to fresh air immediately. If not breathing, give artificial respiration. If breathing is difficult, give oxygen. Get medical aid if cough or other symptoms appear.

Protective Equipment and Precautions for Firefighters
As in any fire, wear self-contained breathing apparatus pressure-demand, MSHA/NIOSH (approved or equivalent) and full protective gear.

Extinguishing Media: Substance is noncombustible; use agent most appropriate to extinguish surrounding fire. Use water fog, dry chemical, carbon dioxide or alcohol type foam.

Sodium acetate (CAS#127-09-3)

Physical State: Solid   Appearance: white   Odor: odorless

NFPA Rating: (estimated) Health: 1; Flammability: 1; Reactivity: 0

EMERGENCY OVERVIEW
May cause eye and skin irritation. Hygroscopic. May cause respiratory tract irritation.

FIRST AID
Eyes: Flush eyes with plenty of water for at least 15 minutes, occasionally lifting the upper and lower eyelids. Get medical aid.
Skin: Get medical aid. Flush skin with plenty of soap and water for at least 15 minutes while removing contaminated clothing and shoes. Wash clothing before reuse.
Ingestion: If victim is conscious and alert, give 2-4 cupfuls of milk or water. Never give anything by mouth to an unconscious person. Get medical aid.
Inhalation: Remove from exposure to fresh air immediately. If not breathing, give artificial respiration. If breathing is difficult, give oxygen. Get medical aid if cough or other symptoms appear.

Protective Equipment and Precautions for Firefighters
As in any fire, wear self-contained breathing apparatus pressure-demand, MSHA/NIOSH (approved or equivalent) and full protective gear.

Extinguishing Media: For small fires, use water spray, dry chemical, carbon dioxide or chemical foam.

Sodium bichromate (see Sodium Dichromate)

Sodium carbonate (CAS#497-19-8)

Physical State: Crystalline powder Appearance: colorless to white Odor:odorless

NFPA Health 2 Flammability 0 Stability 0

EMERGENCY OVERVIEW
Causes eye and skin irritation. Causes respiratory tract irritation. Hygroscopic.

FIRST AID

Eyes: Immediately flush eyes with plenty of water for at least 15 minutes, occasionally lifting the upper and lower eyelids. Get medical aid.
Skin: Get medical aid. Flush skin with plenty of soap and water for at least 15 minutes while removing contaminated clothing and shoes. Wash clothing before reuse.
Ingestion: Do NOT induce vomiting. If victim is conscious and alert, give 2-4 cupfuls of milk or water. Never give anything by mouth to an unconscious person. Get medical aid.
Inhalation: Remove from exposure to fresh air immediately. If not breathing, give artificial respiration. If breathing is difficult, give oxygen. Get medical aid.

Protective Equipment and Precautions for Firefighters
As in any fire, wear self-contained breathing apparatus pressure-demand, MSHA/NIOSH (approved or equivalent) and full protective gear.

Extinguishing Media: Substance is noncombustible; use agent most appropriate to extinguish surrounding fire.

Sodium chromate (CAS#7775-11-3)

Appearance Yellow Physical State Solid odor odorless

NFPA Health 3 Flammability 0 Instability 1

EMERGENCY OVERVIEW
Oxidizer: Contact with combustible/organic material may cause fire. Cancer hazard. May be fatal if inhaled. Toxic if swallowed. Harmful in contact with skin. Causes burns by all exposure routes.

FIRST AID
Eye Contact Rinse immediately with plenty of water, also under the eyelids, for at least 15 minutes. Immediate medical attention is required.
Skin Contact Wash off immediately with plenty of water for at least 15 minutes. Immediate medical attention is required.
Inhalation Move to fresh air. If breathing is difficult, give oxygen. Do not use mouth-to-mouth resuscitation if victim ingested or inhaled the substance; induce artificial respiration with a respiratory medical device. Immediate medical attention is required.
Ingestion Do not induce vomiting. Call a physician or Poison Control Center immediately.

Protective Equipment and Precautions for Firefighters

As in any fire, wear self-contained breathing apparatus pressure-demand, MSHA/NIOSH (approved or equivalent) and full protective gear.

Suitable Extinguishing Media Carbon dioxide (CO_2). Dry chemical. chemical foam.

Sodium dichromate (CAS#10588-01-9)

Physical state Solid. Appearance Red-orange crystals.

NFPA ratings Health: 3 Flammability: 0 Instability: 0

Emergency overview
May be fatal if inhaled. Causes skin and eye burns. Harmful if swallowed or absorbed through skin. Causes severe respiratory tract irritation. Cancer hazard - can cause cancer. May cause allergic respiratory and skin reactions.

FIRST AID
Eye contact Immediately flush with plenty of water for at least 15 minutes. If easy to do, remove contact lenses. Call a physician or poison control center immediately.
Skin contact Immediately flush with plenty of water for at least 15 minutes while removing contaminated clothing and shoes. Call a physician or poison control center immediately. Destroy contaminated clothing and shoes.
Inhalation Move to fresh air. If breathing stops, provide artificial respiration. For breathing difficulties, oxygen may be necessary. Call a physician or poison control center immediately.
Ingestion Call a physician or poison control center immediately. DO NOT induce vomiting. if victim is fully conscious, give a cupful of water. Never give anything by mouth to an unconscious person. If vomiting occurs, keep head lower than the hips to help prevent aspiration.

Protective Equipment and Precautions for Firefighters
As in any fire, wear self-contained breathing apparatus pressure-demand, MSHA/NIOSH (approved or equivalent) and full protective gear.

Suitable extinguishing media
Use appropriate extinguishing media for any nearby fire. Cool containers exposed to heat with

water spray and remove container, if no risk is involved. Containers may explode when heated.

Sodium hydroxide (CAS#1310-73-2)

Physical State: Liquid Appearance: clear Odor: none reported

NFPA Rating: (estimated) Health: 3; Flammability: 0; Instability: 1

EMERGENCY OVERVIEW
Causes eye and skin burns. May cause severe respiratory tract irritation with possible burns. May cause severe digestive tract irritation with possible burns. Corrosive to aluminum. Eye contact may result in permanent eye damage.

FIRST AID
Eyes: In case of contact, immediately flush eyes with plenty of water for at least 15 minutes. Get medical aid immediately.
Skin: In case of contact, immediately flush skin with plenty of water for at least 15 minutes while removing contaminated clothing and shoes. Get medical aid immediately. Wash clothing before reuse.
Ingestion: If swallowed, do NOT induce vomiting. Get medical aid immediately. If victim is fully conscious, give a cupful of water. Never give anything by mouth to an unconscious person.
Inhalation: If inhaled, remove to fresh air. If not breathing, give artificial respiration. If breathing is difficult, give oxygen. Get medical aid.

Protective Equipment and Precautions for Firefighters
As in any fire, wear self-contained breathing apparatus pressure-demand, MSHA/NIOSH (approved or equivalent) and full protective gear.

Extinguishing Media: Use water spray to cool fire-exposed containers. Substance is noncombustible; use agent most appropriate to extinguish surrounding fire.

Sodium persulfate (CAS#7772-27-1)

Physical State: White crystals Odor: None
NFPA Ratings: Health 2 Flammability 0 Reactivity 2

EMERGENCY OVERVIEW

Oxidizer. Greatly increases the burning rate of combustible materials. This material in sufficient quantity and reduced particle size is capable of creating a dust explosion.

FIRST AID
Eyes: Remove contact lenses. Flush with water. Get medical aid if irritation occurs or persist.
Skin: Wash skin with soap and water. Get medical aid if symptoms persist.
Inhalation: Immediately remove from exposure to fresh air. If not breathing, give artificial respiration. If breathing is difficult, give oxygen. Get medical aid.
Ingestion: Do not induce vomiting. Rinse mouth with water. If conscious, give 1-2 glasses of water. Get doctor immediately. Never give anything by mouth to an unconscious person.

Protective Equipment and Precautions for Firefighters
As in any fire, wear self-contained breathing apparatus pressure-demand, MSHA/NIOSH (approved or equivalent) and full protective gear.

Extinguishing Media: Deluge with water.

Sulfuric Acid (CAS#7664-93-9)

Appearance: clear colorless - oily liquid

NFPA Rating: Health 3 Flammability 0 Instability 2 Physical hazards W

EMERGENCY OVERVIEW Corrosive. Causes eye and skin burns. May cause severe respiratory tract irritation with possible burns. May cause severe digestive tract irritation with possible burns. Cancer hazard. May cause fetal effects based upon animal studies. May cause kidney damage. May be fatal if inhaled. May cause lung damage. Hygroscopic. Strong oxidizer. Contact with other material may cause a fire. May cause severe eye, skin and respiratory tract irritation with possible burns.

First Aid
Eyes: Get medical aid immediately. Do NOT allow victim to rub or keep eyes closed. Extensive irrigation with water is required (at least 30 minutes).
Skin: Get medical aid immediately. Immediately flush skin with plenty of soap and water for at least 15 minutes while removing contaminated clothing and shoes. Wash clothing before reuse. Destroy contaminated shoes.
Ingestion: Do NOT induce vomiting. If victim is conscious and alert, give 2-4 cupfuls of milk or water. Never give anything by mouth to an unconscious person. Get medical aid immediately.
Inhalation: Get medical aid immediately. Remove from exposure to fresh air immediately. If breathing is difficult, give oxygen. Do NOT use mouth-to-mouth resuscitation. If breathing has ceased apply artificial respiration using oxygen and a suitable mechanical device such as a bag and a mask.

Protective Equipment and Precautions for Firefighters
As in any fire, wear self-contained breathing apparatus pressure-demand, MSHA/NIOSH (approved or equivalent) and full protective gear.

Suitable Extinguishing Media Substance is nonflammable; use agent most appropriate to extinguish surrounding fire..

DO NOT USE WATER!

Sulphur Dioxide (CAS#7446-09-5)

Colorless gas with a highly irritating, pungent odor.

NFPA Health: 3 Flammability: 0 Reactivity: 0

EMERGENCY OVERVIEW
Corrosive to exposed tissues. Inhalation of vapors may result in pulmonary edema and chemical pneumonitis. Nonflammable. Reacts with water to produce sulfuric acid.

EYES:
PERSONS WITH POTENTIAL EXPOSURE SHOULD NOT WEAR CONTACT LENSES. Flush
contaminated eyes with copious quantities of water. Part eyelids to assure complete flushing. Continue for a minimum of 15 minutes. Seek immediate medical attention.
SKIN:

Remove contaminated clothing as rapidly as possible. Flush affected area with copious quantities of water. Seek immediate medical attention.

INGESTION:

Not required.

INHALATION:

PROMPT MEDICAL ATTENTION IS MANDATORY IN ALL CASES OF OVER EXPOSURE. RESCUEPERSONNEL SHOULD BE EQUIPPED WITH SELF-CONTAINED BREATHING APPARATUS. Victims should be assisted to an uncontaminated area and inhale fresh air. Quick removal from the contaminated area is most important. If breathing has stopped administer artificial resuscitation and supplemental oxygen.

Protective Equipment and Precautions for Firefighters

As in any fire, wear self-contained breathing apparatus pressure-demand, MSHA/NIOSH (approved or equivalent) and full protective gear.

Suitable extinguishing media:

Use media appropriate for surrounding materials. Sulfur dioxide forms sulfuric acid solutions with water.

t-Butyl alcohol (CAS#75-65-0)

Physical State: Liquid Appearance: after melting, clear colorless

Odor: camphor

NFPA Rating: health-1; flammability-3; reactivity-0

EMERGENCY OVERVIEW

Flammable liquid May cause central nervous system depression. May be absorbed throughthe skin. May cause eye and skin irritation. May cause respiratoryand digestive tract irritation.

Target Organs: Kidneys, central nervous system, liver.

First Aid

Eyes: Immediately flush eyes with plenty of water for at least 15 minutes, occasionally lifting the upper and lower lids. Get medical aid. Do NOT allow victim to rub or keep eyes closed.
Skin: Flush skin with plenty of soap and water for at least 15 minutes while removing contaminated clothing and shoes. Get medical aid if irritation develops or persists.
Ingestion: Do NOT induce vomiting. If victim is conscious and alert, give 2-4 cupfuls of milk or water. Get medical aid immediately. Do NOT induce vomiting. Allow the victim to rinse his mouth and then to drink 2-4 cupfuls of water, and seek medical advice.
Inhalation: Remove from exposure to fresh air immediately. If not breathing, give artificial respiration. If breathing is difficult, give oxygen. Get medical aid.

Protective Equipment and Precautions for Firefighters
As in any fire, wear self-contained breathing apparatus pressure-demand, MSHA/NIOSH (approved or equivalent) and full protective gear.

Extinguishing Media: For small fires, use dry chemical, carbon dioxide, water spray or alcohol-resistant foam. For large fires, use water spray, fog, or alcohol-resistant foam. In case of fire, use water, dry chemical, chemical foam, or alcohol-resistant foam.

Toluene (CAS#108-88-3)

Physical State: Liquid Appearance: colorless Odor: sweetish, pleasant odor - benzene-like

NFPA Rating: (estimated) Health: 2; Flammability: 3; Instability: 0

EMERGENCY OVERVIEW
Flammable liquid and vapor. Causes eye, skin, and respiratory tract irritation. Breathing vapors may cause drowsiness and dizziness. May be absorbed through intact skin. Aspiration hazard if swallowed. Can enter lungs and cause damage.

First Aid
Eyes: In case of contact, immediately flush eyes with plenty of water for a t least 15 minutes. Get medical aid.
Skin: In case of contact, flush skin with plenty of water. Remove contaminated clothing and shoes. Get medical aid if irritation develops and persists. Wash clothing before reuse.

Ingestion: Potential for aspiration if swallowed. Get medical aid immediately. Do not induce vomiting unless directed to do so by medical personnel. Never give anything by mouth to an unconscious person. If vomiting occurs naturally, have victim lean forward.
Inhalation: If inhaled, remove to fresh air. If not breathing, give artificial respiration. If breathing is difficult, give oxygen. Get medical aid.

Protective Equipment and Precautions for Firefighters
As in any fire, wear self-contained breathing apparatus pressure-demand, MSHA/NIOSH (approved or equivalent) and full protective gear.

Extinguishing Media: Use water spray, dry chemical, carbon dioxide, or appropriate foam. Solid streams of water may be ineffective and spread material.

Tributylphosphate (CAS#126-73-8)

Odorless colorless to yellow liquid

NFPA Health 3 Flammability 1 Instability 0

EMERGENCY OVERVIEW
Potential symptoms of overexposure are irritation of eyes, respiratory system and skin...

First Aid
Skin Get medical aid. Immediately flush skin with plenty of soap and water for at least 15 minutes while removing contaminated clothing/shoes.
Inhalation Remove person to fresh air. If signs/symptoms continue, get medical attention. Give oxygen or artificial respiration as needed.
Eyes Thoroughly flush the eyes with large amounts of clean low-pressure water for at least 15 minutes, occasionally lifting the upper and lower eyelids. Seek medical attention.
Ingestion NEVER give anything by mouth to an unconscious person. If vomiting does occur, have victim lean forward to prevent aspiration. Rinse mouth with water. Get medical attention.

Protective Equipment and Precautions for Firefighters

As in any fire, wear self-contained breathing apparatus pressure-demand, MSHA/NIOSH (approved or equivalent) and full protective gear.

Suitable extinguishing media:
Use water spray, alcohol-resistant foam, dry chemical or carbon dioxide.

Turpentine (CAS#8006-64-2)

Liquid, Clear to pale amber   Odor Sulfur-terpene.

NFPA ratings Health: 2   Flammability: 3   Instability: 1

Emergency overview Flammable liquid and vapor. Harmful or fatal if swallowed. Exposure may occur by inhalation, ingestion and through the skin. May cause skin, eye, and respiratory tract irritation and/or injury.

First Aid
Eye contact In case of eye irritation rinse immediately with plenty of water, also under the eyelids for at least 15 minutes. Obtain medical attention if irritation persists.
Skin contact Remove and isolate contaminated clothing and shoes. Immediately flush skin with plenty of water. Wash with soap and water. Launder contaminated clothing before reuse. Get medical attention if irritation develops or persists.
Inhalation Rescuers should put on appropriate protective gear. Remove victim from area of exposure. If not breathing, give artificial respiration or give oxygen by trained personnel. Call a physician or poison control center immediately.
Ingestion Call a physician or poison control center immediately.

Protective Equipment and Precautions for Firefighters
As in any fire, wear self-contained breathing apparatus pressure-demand, MSHA/NIOSH (approved or equivalent) and full protective gear.

Suitable extinguishing media
Dry chemical, carbon dioxide or alcohol foam. Use water spray to cool / or disperse vapors.

Xylene (CAS#1330-20-7)

Clear, colorless liquid.

NFPA Rating: (estimated) Health: 2; Flammability: 3; Instability: 0

EMERGENCY OVERVIEW
Flammable liquid and vapor. Aspiration hazard if swallowed.

First Aid
Eyes: In case of contact, immediately flush eyes with plenty of water for at least 15 minutes. Get medical aid.
Skin: In case of contact, flush skin with plenty of water. Remove contaminated clothing and shoes. Get medical aid if irritation develops and persists. Wash clothing before reuse.
Ingestion: Potential for aspiration if swallowed. Get medical aid immediately. Do not induce vomiting unless directed to do so by medical personnel. Never give anything by mouth to an unconscious person.
Inhalation: If inhaled, remove to fresh air. If not breathing, give artificial respiration. If breathing is difficult, give oxygen. Get medical aid.

Protective Equipment and Precautions for Firefighters
As in any fire, wear self-contained breathing apparatus pressure-demand, MSHA/NIOSH (approved or equivalent) and full protective gear.

Extinguishing Media: Use water spray to cool fire-exposed containers. Water may be ineffective. This material is lighter than water and insoluble in water. The fire could easily be spread by the use of water in an area where the water cannot be contained. Use water spray, dry chemical, carbon dioxide, or appropriate foam.

Zinc bromide (CAS#7699-45-8)

Appearance: Granular powder. Odor: Odorless.

NFPA Health 3 **Flammability** 0 **Instability** 0

Hazards Stable under ordinary conditions of use and storage.

First Aid

Inhalation: Remove to fresh air. If not breathing, give artificial respiration. If breathing is difficult, give oxygen. Get medical attention immediately.
Ingestion: If swallowed, give several glasses of water to drink. Vomiting may occur spontaneously, but DO NOT INDUCE! Never give anything by mouth to an unconscious person. Get medical attention.
Skin Contact: Wipe off excess material from skin then immediately flush skin with plenty of water for at least 15 minutes while removing contaminated clothing and shoes. Get medical attention immediately. Wash clothing before reuse. Thoroughly clean shoes before reuse.
Eye Contact: Immediately flush eyes with plenty of water for at least 15 minutes, lifting lower and upper eyelids occasionally. Get medical attention immediately.

Protective Equipment and Precautions for Firefighters
As in any fire, wear self-contained breathing apparatus pressure-demand, MSHA/NIOSH (approved or equivalent) and full protective gear.

Fire: Not considered to be a fire hazard. Finely divided zinc compounds can explode in air. Specific levels have not been described.
Explosion: Not considered to be an explosion hazard.
Fire Extinguishing Media: Use any means suitable for extinguishing surrounding fire. Use water carefully as material will react with water to form acidic solution.

Zinc chloride (CAS#7646-85-7)

White Odorless crystalline granules.

EMERGENCY OVERVIEW
Corrosive. Contact with skin causes irritation and possible burns, especially if the skin is wet or moist. Harmful if swallowed. May cause severe respiratory and digestive tract irritation with possible burns. Toxic. Causes severe eye irritation and possible injury.

First Aid

Inhalation: Remove to fresh air. If not breathing, give artificial respiration. If breathing is difficult, give oxygen. Get medical attention immediately.
Ingestion: If swallowed, DO NOT INDUCE VOMITING. Give large quantities of water. Never give anything by mouth to an unconscious person. Get medical attention immediately.
Skin Contact: Immediately flush skin with plenty of water for at least 15 minutes while removing contaminated clothing and shoes. Get medical attention immediately. Wash clothing before reuse. Thoroughly clean shoes before reuse.
Eye Contact: Immediately flush eyes with plenty of water for at least 15 minutes, lifting lower and upper eyelids occasionally. Get medical attention immediately.

Protective Equipment and Precautions for Firefighters
As in any fire, wear self-contained breathing apparatus pressure-demand, MSHA/NIOSH (approved or equivalent) and full protective gear.

Fire Fighting Measures

Fire: Not considered to be a fire hazard.
Explosion: Not considered to be an explosion hazard.
Fire Extinguishing Media: Use any means suitable for extinguishing surrounding fire.

Note: To the best of my knowledge, the information contained herein is accurate. However, the publisher assumes no liability whatsoever for the accuracy or completeness of the information contained herein. Final determination of suitability of any material is the sole responsibility of the user. All materials may present unknown hazards and should be used with caution. Although certain hazards are described herein, I cannot guarantee that these are the only hazards that exist. Not all events are survivable.

NFPA 704 DESCRIPTION

"NFPA 704: Standard System for the Identification of the Hazards of Materials for Emergency Response"

Codes

The four divisions are typically color-coded:

RED = flammability

BLUE= Level of health hazard

YELLOW = Chemical reactivity

WHITE = Codes for special hazards.

Each of health, flammability and reactivity is rated on a scale from 0 (no hazard) to 4 (severe risk).

Flammability (red)	
0	Materials that will not burn under typical fire conditions (e.g. carbon dioxide), including intrinsically noncombustible materials such as concrete, stone and sand (Materials that will not burn in air when exposed to a temperature of 820 °C (1,500 °F) for a period of 5 minutes)
1	Materials that require considerable preheating, under all ambient temperature conditions, before ignition and combustion can occur (e.g. mineral oil). Includes some finely divided suspended solids that do not require heating before ignition can occur. Flash point at or above 93 °C (200 °F).
2	Must be moderately heated or exposed to relatively high ambient temperature before ignition can occur (e.g. diesel fuel) and some finely divided suspended solids that do not require heating before ignition can occur. Flash point between 38 and 93 °C (100 and 200 °F).

3	Liquids and solids (including finely divided suspended solids) that can be ignited under almost all ambient temperature conditions (e.g. gasoline). Liquids having a flash point below 23 °C (73 °F) and having a boiling point at or above 38 °C (100 °F) or having a flash point between 23 and 38 °C (73 and 100 °F).
4	Will rapidly or completely vaporize at normal atmospheric pressure and temperature, or is readily dispersed in air and will burn readily (e.g. acetylene, diethylzinc). Includes pyrophoric substances. Flash point below 23 °C (73 °F).
Health (blue)	
0	Poses no health hazard, no precautions necessary and would offer no hazard beyond that of ordinary combustible materials (e.g. wood)
1	Exposure would cause irritation with only minor residual injury (e.g. acetone)
2	Intense or continued but not chronic exposure could cause temporary incapacitation or possible residual injury (e.g. diethyl ether)
3	Short exposure could cause serious temporary or moderate residual injury (e.g. chlorine)
4	Very short exposure could cause death or major residual injury (e.g. hydrogen cyanide, phosphine, carbon monoxide, sarin, hydrofluoric acid)
Instability/Reactivity (yellow)	
0	Normally stable, even under fire exposure conditions, and is not reactive with water (e.g. helium)
1	Normally stable, but can become unstable at elevated temperatures and pressures (e.g. propene)
2	Undergoes violent chemical change at elevated temperatures and pressures, reacts violently with water, or may form explosive mixtures with water (e.g. white phosphorus, potassium, sodium)

<table>
<tr><td>3</td><td>Capable of detonation or explosive decomposition but requires a strong initiating source, must be heated under confinement before initiation, reacts explosively with water, or will detonate if severely shocked (e.g. ammonium nitrate, chlorine trifluoride)</td></tr>
<tr><td>4</td><td>Readily capable of detonation or explosive decomposition at normal temperatures and pressures
(e.g.nitroglycerin, chlorine azide, chlorine dioxide)</td></tr>
<tr><td colspan="2">Special notice (white)</td></tr>
<tr><td colspan="2">The white "special notice" area can contain several symbols. The following symbols are defined by the NFPA 704 standard.</td></tr>
<tr><td>OX</td><td>Oxidizer, allows chemicals to burn without an air supply (e.g. potassium perchlorate, ammonium nitrate, hydrogen peroxide).</td></tr>
<tr><td>~~W~~</td><td>Reacts with water in an unusual or dangerous manner (e.g. cesium,sodium, sulfuric acid).</td></tr>
<tr><td>SA</td><td>Simple asphyxiant gas. Specifically limited to the following gases: nitrogen,helium, neon, argon, krypton and xenon.</td></tr>
<tr><td colspan="2">Non-standard symbols (white)</td></tr>
<tr><td colspan="2">These hazard codes are not part of the NFPA 704 standard, but are occasionally used in an unofficial manner. The use of non-standard codes may be permitted, required or disallowed by the authority having jurisdiction (e.g. fire department).</td></tr>
<tr><td>COR
ACID, ALK</td><td>Corrosive; strong acid or base (e.g. sulfuric acid, potassium hydroxide)
Acid or alkaline, to be more specific</td></tr>
<tr><td>BIO or ☣</td><td>Biological hazard (e.g. smallpox virus)</td></tr>
<tr><td>POI</td><td>Poisonous (e.g. strychnine)</td></tr>
<tr><td>RA, RAD or ☢</td><td>Radioactive (e.g. plutonium, uranium)</td></tr>
<tr><td>CYL or CRYO</td><td>Cryogenic (e.g. liquid nitrogen)</td></tr>
</table>

HAZMAT GLOSSARY

Absorption The passing of a substance into the circulatory system of the body. Also used specifically to refer to entry of toxicants through the skin.
Acute Exposure An exposure to a toxic substance which occurs in a short or
single time period.
Acute Toxicity Any poisonous effect produced by a single short-term exposure. The LD50 of a substance (the lethal dose at which 50 percent of test animals succumb to the toxicity of the chemicals) is typically used as a measure of its acute toxicity.
Additive Effect A biological response to exposure to multiple chemicals which is equal to the sum of the effects of the individual agents.
Adsorption The bonding of chemicals to soil particles or other surfaces.
Aerosol A solid particle or liquid droplet suspended in air. An aerosol is larger than a molecule and can be filtered from the air.
Antagonism The situation in which two chemicals interfere with each other's actions, or one chemical interferes with the action of the other.
Aquifer An underground bed, or layer, of earth, gravel, or porous storage that contains water.
Asphyxiants Chemicals that starve the cells of an individual from the life-giving oxygen needed to sustain metabolism.
Biodegradable Capable of decomposing quickly through the action of microorganisms.
Biomagnification The tendency of certain chemicals to become concentrated as they move into and up the food chain.
Boiling Point The temperature at which a liquid will start to become a gas, and boil. A chemical with a low boiling point can boil and evaporate quickly. If a material that is flammable also has a low boiling point, a special fire hazard exists.
Carcinogen A chemical or physical agent that encourages cells to develop cancer.
Central Nervous System Depressants Toxicants that deaden the central nervous system (CNS), diminishing sensation.
CHEMTREC Chemical Transportation Emergency Center, a service operated by the Chemical Manufacturers Association to provide information and other assistance to emergency responders.

Chronic Exposure Process by which small amounts of toxic substances are taken into the body over an extended period.
Command Post A centralized base of operations established near the site of a hazardous materials incident.
Corrosive A chemical that destroys or irreversibly alters living tissue by direct chemical action at the site of contact.
Decontamination The process of removing or neutralizing contaminants that have accumulated on personnel and equipment. This process is critical to health and safety at hazardous waste incidents.
Dermal Exposure Exposure to toxic substances by entry though the skin.
Dose The quantity of a chemical absorbed and available for interaction with metabolic processes.
Epidemiology Studies Investigation of factors contributing to disease or toxic effects in the general population.
Evaporation Rate The rate at which a chemical changes into a vapor. A chemical that evaporates quickly can be a more dangerous fire or health hazard.
Exercise A simulated emergency condition carried out for the purpose of testing and evaluating the readiness of a community or organization to handle a particular type of emergency.
Explosive A chemical that causes a sudden, almost instantaneous release of pressure, gas, and heat when subjected to sudden shock, pressure, or high temperatures.
Extremely Hazardous Any one of more than 300 hazardous chemicals on a list Substance (EHS) compiled by EPA to provide a focus for State and local emergency planning activities.
Hazard Class A group of materials, as designated by the Department of Transportation, that share a common major hazardous property such as radioactivity or flammability.
Hazardous Materials Response Team (HMRT) A team of specially trained personnel who respond to hazardous materials incident. The team performs various response actions including assessment, firefighting, rescue, and containment; they are not responsible for cleanup operations following the incident.
Incident Commander The person in charge of on-scene coordination of a response to an incident, usually a senior officer in a fire department.
Inversion An atmospheric condition caused by a layer of warm air preventing cool air trapped beneath it from rising, thus holding down pollutants that could otherwise be dispersed.
Irritant Chemicals which inflame living tissue by chemical action at the site of contact, causing pain or swelling.

LD50 The calculated dosage of a material that would be fatal to 50% of an exposed population (Lethal Dose 50%).
Leachate Material that pollutes water as it seeps through solid waste.
Leaching The process by which water dissolves nutrient chemicals or contaminants and carries them away, or moves them to a lower layer.
LEPC Local Emergency Planning Committee.
LOAEL The Lowest Observed Adverse Effect Level, i.e., the lowest dose which produces an observable adverse effect.
Medium The environmental vehicle by which a pollutant is carried to the receptor (e.g., air, surface water, soil, or groundwater).
Melting Point The temperature at which a solid material changes to a liquid. Solid materials with low melting points should not be stored in hot areas.
Mg Milligram, a metric unit of mass, one thousandth of a gram: 1 mg = 0.001 g = 1000 µg.
Mm3 Milligrams per cubic meter. The mass in micrograms of a substance contained within a cubic meter of another substance or vacuum. This is the standard unit of measure for the mass density (concentration) of particles suspended in air; also sometimes used for the concentration of gases in air.
MSDS (Material Safety Data Sheet) A worksheet required by the U.S. Occupational Safety and Health Administration (OSHA) containing information about hazardous chemicals in the workplace. MSDSs are used to fulfill part of the hazardous chemical inventory reporting requirements under the Emergency Planning and Community Right-to-Know Act.
Mutagen A chemical or physical agent that induces a permanent change in the genetic material.
NOAEL No Observable Adverse Effect Level.
NECP Suit Non-encapsulating chemical protective suit. Not gas or vapor tight.
Organic Compound Chemicals that contain carbon. Volatile organic compounds vaporize at room temperature and pressure. They are found in many indoor sources, including many common household products and building materials.
Pathway A history of the flow of a pollutant from source to receptor, including qualitative descriptions of emission type, transport, medium, and exposure route.
PEL Permissible Exposure Limits set by OSHA as a guide to acceptable levels of chemical exposure.

Percent Volatile The percentage of a chemical that will evaporate at ordinary temperatures. A high volatile percentage may mean there is more risk of explosion, or that dangerous fumes can be released. Evaporation rates are a better measure of the danger than the percent volatile measure.

PH The pH is a measure of how acidic or caustic a chemical is, based on a scale of 1 to 14. A pH of 1 means the chemical is very acidic. Pure water has a pH of 7. A pH of 14 means the chemical is very caustic. Both acidic and caustic substances are dangerous to skin and other valuable surfaces.

Poison A chemical that, in relatively small amounts, is able to produce injury by chemical action when it comes in contact with a susceptible tissue.

RCRA The Resource Conservation and Recovery Act (of 1976). A Federal statute which establishes a framework for proper management and disposal of all wastes. Generation, transportation, storage, treatment, and disposal of hazardous wastes are all regulated under this Act.

Risk Assessment Broadly defined as the scientific activity of evaluating the toxic properties of a chemical and the conditions of human exposure to it, with the objective of determining the probability that exposed humans will be adversely affected. Its four main components are:

1. Hazard Identification—Does the agent cause the effect?
2. Dose-Response Assessment—What is the relationship between the dose and its incidence in human beings?
3. Exposure Assessment—What exposures are experienced or anticipated, and under what conditions?
4. Risk Characterization—The total analysis producing an estimate of the incidence of the adverse effect in a given population.

Runoff Water from rain, snow melt, or irrigation that flows over the ground surface and returns to streams.

SARA Superfund Amendments and Reauthorization Act of 1986.

SERC State Emergency Response Commission.

Solubility in Water An indicator of the amount of a chemical that can be dissolved in water, shown as a percentage or as a description. A low percent of solubility (or a description of "slight" solubility or "low" solubility) means that only a small amount will dissolve in water. Knowing this may help firefighters or personnel cleaning a spill.

Specific Gravity A comparison of the weight of the chemical to the weight of an equal volume of water. Chemicals with a specific gravity of less than 1 are lighter than water, while a specific gravity of more than 1 means the chemical is heavier than water. Most flammable liquids are lighter than water.

Synergistic Effect A biological response to exposure to multiple chemicals which is greater than the sum of the effects of the individual agents.

Systemic Toxicants Chemical compounds that affect entire organ systems, often operating far from the original site of entry.

Teratogen A material that produces a physical defect in a developing embryo.

Threshold The lowest dose of a chemical at which a specific measurable effect is observed. Below this dose, the effect is not observed.

Title III The third part of SARA, also known as the Emergency Planning and Community Right-to-Know Act of 1986.

TLV Threshold Limit Values, which are the calculated airborne concentrations of a substance to which all workers could be repeatedly exposed 8 hours a day without adverse effects.

TECP Suit Totally encapsulating chemical protective suit. Special protective suits made of material that prevents toxic or corrosive substances or vapors from coming in contact with the body. Gas and vapor tight suit.

Toxicity The degree of danger posed by a substance to animal or plant life.

Toxicology The study of the adverse effects of chemicals on biological systems, and the assessment of the probability of their occurrence.

Transformation The chemical alteration of a compound by processes such as reaction with other compounds or breakdown into component elements.

Transport Hydrological, atmospheric, or other physical processes that convey pollutants through and across media from source to receptor.

Vapor Density The measure of the heaviness of a chemical's vapor as compared to the weight of a similar amount of air. A vapor density of 1.0 is equal to air. Vapors that are heavier than air may build up in low-lying areas, such as along floors, in sewers, or in elevator shafts. Vapors that are lighter than air rise and may collect near the ceiling.

Vapor Pressure The measure of how quickly a chemical liquid will evaporate. Chemicals with low boiling points have high vapor pressures. If a chemical with a high vapor pressure spills, there is an increased risk of explosion and a greater risk that workers will inhale toxic fumes.

Volatilization Entry of contaminants into the atmosphere by evaporation from soil or water.

HYDROGEN SULFIDE H2S

- ! Disclaimer: this information is advisory in nature and is not intended to identify all scenarios or situations a person might encounter.
- ! Following these guidelines will not guarantee your safety.

	*WARNING! . High concentrations of H2S can cause immediate collapse with loss of breathing, even after inhalation of a **single breath**.*

Other Names: Sewer gas, Stink damp, manure gas, swamp gas, Sulfane, Dihydrogen monosulfide, Dihydrogen sulfide, Sulfurated hydrogen, Sulfureted hydrogen, Sulfuretted hydrogen, Sulfur hydride, Hydrosulfuric acid.

Hydrogen sulfide is a colorless, flammable, extremely hazardous gas with a "rotten egg" smell.

Workers in oil and natural gas drilling and refining may be exposed because hydrogen sulfide may be present in oil and gas deposits and is a by-product of the desulfurization process of these fuels.
It occurs naturally in crude petroleum, natural gas, and hot springs. In addition, hydrogen sulfide is produced by bacterial breakdown of organic materials and human and animal wastes (e.g., sewage). Industrial activities that can produce the gas include petroleum /natural gas drilling and refining, mining, wastewater treatment, paving, coke ovens, tanneries, and kraft paper mills. Hydrogen sulfide can also exist as a liquid compressed gas.

Hazardous properties of H_2S gas
Hydrogen sulfide is heavier than air and may travel along the ground. It collects in low-lying and enclosed, poorly-ventilated areas such as basements, manholes, sewer lines, underground telephone vaults and manure pits.
For work within confined spaces, use appropriate procedures for identifying hazards, monitoring and entering confined spaces.

The primary route of exposure is inhalation and the gas is rapidly absorbed by the lungs. Absorption through the skin is minimal. People can smell the "rotten egg" odor of hydrogen sulfide at low concentrations in air. However, with continuous low-level exposure, or at high concentrations, a person loses his/her ability to smell the gas even though it is still present (olfactory fatigue). This can happen very rapidly and at high concentrations, the ability to smell the gas can be lost instantaneously. Therefore, DO NOT rely on your sense of smell to indicate the continuing presence of hydrogen sulfide or to warn of hazardous concentrations. **Lethal exposure to hydrogen sulfide around sour gas wells could occur as far as 2000 meters from the source.**

In addition, hydrogen sulfide is a highly flammable gas and gas/air mixtures can be explosive. It may travel to sources of ignition and flash back. If ignited, the gas burns to produce toxic vapors and gases, such as sulfur dioxide.

Contact with liquid hydrogen sulfide causes frostbite. If clothing becomes wet with the liquid, avoid ignition sources, remove the clothing and isolate it in a safe area to allow the liquid to evaporate.

Health effects of H_2S exposure

Hydrogen sulfide is both an irritant and a chemical asphyxiant with effects on both oxygen utilization and the central nervous system. Its health effects can vary depending on the level and duration of exposure. Repeated exposure can result in health effects occurring at levels that were previously tolerated without any effect.

Low concentrations irritate the eyes, nose, throat and respiratory system (e.g., burning/ tearing of eyes, cough, shortness of breath). Asthmatics may experience breathing difficulties. The effects can be delayed for several hours, or sometimes several days, when working in low-level concentrations. Repeated or prolonged exposures may cause eye inflammation, headache, fatigue, irritability, insomnia, digestive disturbances and weight loss.

Moderate concentrations can cause more severe eye and respiratory irritation (including coughing, difficulty breathing, accumulation of fluid in the lungs), headache, dizziness, nausea, vomiting, staggering and excitability.

Concentration (ppm) **Symptoms/Effects**

0.00011-0.00033 Typical background concentrations

0.01-1.5 Odor threshold (when rotten egg smell is first noticeable to some). Odor becomes more offensive at 3-5 ppm. Above 30 ppm, odor described as sweet or sickeningly sweet.

2-5 Prolonged exposure may cause nausea, tearing of the

eyes, headaches or loss of sleep. Airway problems (bronchial constriction) in some asthma patients.
20 Possible fatigue, loss of appetite, headache, irritability, poor memory, dizziness.

50-100 Slight conjunctivitis ("gas eye") and respiratory tract irritation after 1 hour. May cause digestive upset and loss of appetite.

100 Coughing, eye irritation, loss of smell after 2-15 minutes (olfactory fatigue). Altered breathing, drowsiness after 15-30 minutes. Throat irritation after 1 hour. Gradual increase in severity of symptoms over several hours. Death may occur after 48 hours.

100-150 Loss of smell (olfactory fatigue or paralysis).

200-300 Marked conjunctivitis and respiratory tract irritation after 1 hour. Pulmonary edema may occur from prolonged exposure.

500-700 Staggering, collapse in 5 minutes. Serious damage to the eyes in 30 minutes. Death after 30-60 minutes.

700-1000 Rapid unconsciousness, "knockdown" or immediate collapse within 1 to 2 breaths, breathing stops, death within minutes.

1000-2000 Nearly instant death

In general, working in the following areas and conditions increases a worker's risk of overexposure to hydrogen sulfide:

Confined spaces (for example pits, manholes, tunnels, wells) where hydrogen sulfide can build up to dangerous levels.

Windless or low-lying areas that increase the potential for pockets of hydrogen sulfide to form.

Marshy landscapes where bacteria break down organic matter to form hydrogen sulfide.

Hot weather that speeds up rotting of manure and other organic materials, and increases the hydrogen sulfide vapor pressure.

All personnel working in an area where concentrations of Hydrogen Sulfide may exceed the 10 Parts Per Million (PPM) should be provided with training before beginning work assignments.

Evaluate Exposure

Identify processes that could release or produce hydrogen sulfide. This includes identifying known sources of hydrogen sulfide and evaluating possible fire and explosion hazards. Use a Process or Job Hazard Analysis for identifying and controlling hazards

Test (monitor) the air for hydrogen sulfide. This must be done by a qualified person. Use the right test equipment, such as an electronic meter that detects hydrogen sulfide gas. Conduct air monitoring prior to and at regular times during any work activity where hydrogen sulfide exposure is possible. When working in confined spaces air monitoring must be conducted in accord with the applicable OSHA standards. Detector tubes, direct reading gas monitors, alarm only gas monitors, and explosion meters are examples of monitoring equipment that may be used to test permit space atmospheres.

DO NOT rely on your sense of smell to indicate the continuing presence of hydrogen sulfide or to warn of harmful levels. You can smell the "rotten egg" odor of hydrogen sulfide at low concentrations in air. But after a while, you lose the ability to smell the gas even though it is still present (olfactory fatigue). This loss of smell can happen very rapidly and at high concentrations and the ability to smell the gas can be lost instantly (olfactory paralysis).

Control Exposures

Use exhaust and ventilation systems to reduce hydrogen sulfide levels. Make sure that the system is:

- Non-sparking
- Grounded
- Corrosion-resistant
- Separate from other exhaust ventilation systems
- Explosion-proof

These safety measures are important because hydrogen sulfide is flammable and can corrode materials if they are not properly protected. When working in confined spaces ventilation should operate continuously and must be conducted in accord with the applicable OSHA standards.

Establish proper rescue procedures to safely rescue someone from a hydrogen sulfide exposure.

WARNING: First responders/ rescuers must be trained and properly protected before entering areas with elevated levels of hydrogen sulfide.

Rescuer protection should include:

- Positive-pressure, self-contained breathing apparatus (SCBA).
- A safety line to allow for rapid exit if conditions become dangerous.

Use respiratory and other personal protective equipment.

Respiratory protection should be at least:

- For exposures below 100 ppm, use an air-purifying respirator with specialized canisters/cartridges for hydrogen sulfide. A full face respirator will provide eye protection.
- For exposures at or above 100 ppm, use a full face pressure demand self-contained breathing apparatus (SCBA) with a minimum service life of thirty minutes or a combination full face pressure demand supplied-air respirator with an auxiliary self-contained air supply.

Exposures at or above 100 ppm are considered immediately dangerous to life and health (IDLH).
Follow OSHA requirements for confined space entry.

Enter the space only if necessary and follow established procedures:

Test (monitor) the air in the space from the outside before entering.

Test (monitor) the air in the space continuously during work operation.

Determine if entry permit is required.

Ventilate area continuously to remove accumulated hydrogen sulfide.

Make sure that rescue procedures, personnel, and equipment (e.g., positive pressure SCBAs) are in place.

Maintain contact with trained attendant.

FIRST AID FOR H2S POISONING

Do Not Attempt to Make Rescue If You Are Not Trained or Lack Proper PPE!

Rescuers should be trained in H2S specific rescue and will require the use of supplied breathing air. (Positive pressure SCBA, Positive pressure SBA).

Remove the victim to fresh air immediately, **only** if it can be done without creating more victims.
INHALATION: Remove to uncontaminated area. Give artificial respiration if not breathing. If breathing is difficult, oxygen should be administered by qualified personnel. **Get immediate medical attention.**

SKIN CONTACT: Wash skin with soap and water for at least 15 minutes while removing contaminated clothing and shoes. Get medical attention, if needed. Thoroughly clean and dry contaminated clothing and shoes before reuse.

EYE CONTACT: Flush eyes with plenty of water for at least 15 minutes. Then get immediate medical attention.

INGESTION: ingestion of a gas is unlikely

Standard Operating Procedure for cleaning up blood or bodily fluids.

Blood, vomit and feces may contain germs that can cause serious infections. People who clean blood and other bodily fluids should reduce the risk of infection to themselves and others by following these procedures:

Procedure for Blood Spills/Vomit/Feces

1. Wear appropriate personal protective equipment, such as disposable gloves when cleaning up a spill. If the possibility of splashing exists, protective eyewear and a gown should be worn. Eye glasses are not considered to be protective eyewear.

2. Dispose with care, any broken glass or sharps (any item having corners, edges, or projections capable of cutting or piercing the skin) into a puncture-proof container. If available, disposal of sharps into an approved sharps container for biomedical waste is preferred.

3. Clean the spill area with paper towel to remove most of the spill. Disinfectants cannot work properly if the surface has blood or other bodily fluids on it. Cloth towels should not be used unless they are to be thrown out.

4. Discard the paper towel soaked with the blood, vomit, feces or fluid in a plastic-lined garbage bin.

5. Care must be taken to avoid splashing or spraying during the clean up process.

6. Clean the affected area with soap and water then disinfect with a 1:10 bleach solution for 10 minutes or an appropriate disinfectant with proven effectiveness against non-enveloped viruses (eg. Poliovirus, Norovirus, Rotavirus). Refer to the manufacturer's label to ensure the disinfectant is left on the contaminated surface for the correct contact time. With bleach, this would mean the surface stays wet for at least 10 minutes.

7. Ventilate the room well when using a bleach solution. Make sure it is not mixed with other cleaning agents.

8. Wipe treated area with paper towels soaked in tap water. Allow the area to dry.

9. Discard contaminated paper towels, gloves and other disposable equipment in a plastic lined garbage bin. Immediately tie and place with regular trash. Take care not to contaminate other surfaces during this process. Change gloves if needed.

10. Practice hand hygiene, either with soap and water or an alcohol-based hand rub of at least 60% concentration, for 15 seconds after gloves are removed. If the hands are visibly soiled, then soap and water should be used over a hand rub.

11. If an injury occurs during the cleaning process, such as a skin puncture with a blood contaminated sharp object, seek medical attention immediately

Mixing a 1:10 Bleach Solution
100 mL bleach: 900 mL of water
(1 cup of bleach: 9 cups of water).
Contact time on surface is 10 minutes

FIRST AID

! Disclaimer This chapter is designed for use when emergency medical care may not be immediately available due to location, weather, disasters, terrorism or other factors.

! Disclaimer: The information in this book is not intended as a substitute for professional medical advice, emergency medical treatment or formal first-aid training. Do not use this information to diagnose or develop a treatment plan for a medical problem or disease without consulting a qualified health care provider. If you or someone near you is in a life-threatening or emergency medical situation, call for or seek professional emergency medical assistance immediately.

Administering First Aid

This chapter is designed for use when emergency medical care may not be immediately available due to location, weather, disasters, terrorism or other factors.

- Always practice universal precautions: wear gloves and other personal protective equipment.
- Always make sure the area is safe to enter or provide first aid.
- Become familiar with the recovery position as it is useful is a number of first aid situations.

Recovery Position

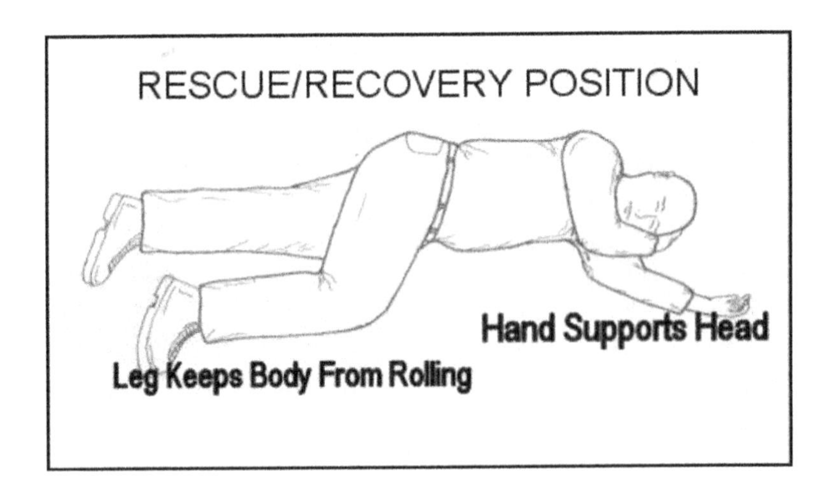

The ABC's of First Aid
In any first aid situation, always think ABC first. ABC—Airway, Breathing, and Circulation are the three most critical basic body functions necessary for life. This holds true for victims of trauma, poisoning, infection, and every other type of injury or illness. ABC can often be assessed quickly and easily, however sometimes a thorough survey is essential.

A and B: Airway and Breathing
If a person is conscious, assessment for an open airway is usually straightforward

- Speaking, crying, and screaming all correlate with an open airway. A person with a blocked airway is not able to speak.
- An unconscious person requires closer evaluation.
- Place your ear close to the person's mouth and listen for breath sounds. Feel for air on your cheek. Look for chest rising and falling with breaths.

C: Circulation
Circulation refers to the heart's ability to effectively pump blood to the rest of the body. Signs of circulation include pulse, movement, normal skin color and warmth.
If any of the ABCs are compromised, begin CPR immediately. See the section on CPR for details, page 107.

Do survey for other injuries such as neck and spine injuries and treat accordingly.

Abdominal Injuries

Abdominal injuries related to trauma can be obvious with open wounds and severe symptoms, however critical injuries can be present in a victim that appears normal or near-normal.

Signs and symptoms of internal abdominal injuries after trauma:

- Pain
- Vomiting, hematemesis (vomiting blood)
- Loss of bowel control, rectal bleeding
- Rapid pulse, low blood pressure
- Abdomen hard, bloated, or discolored

Victims of trauma (even minor trauma) with these symptoms should be evaluated by emergency medical services immediately.

Treating Abdominal Injuries

- First, ensure safe surroundings
- Call for emergency medical services immediately
- Support the victim by placing him in the recovery position, page 86. Do not give any food or liquids to him.
- For open abdominal wounds:
- Try to keep the victim and bystanders calm
- Use gloves and personal protective equipment if available.
- Try to control bleeding with direct pressure to bleeding site.
- If there are protruding intestines or other organs, do not attempt to push them back into the body. Cover the organs with moist gauze or cloth. Do not touch or put pressure on protruding organs.

! WARNING: Continually monitor the victims ABCs and provide appropriate first aid until emergency services arrive.

Allergic Reactions / Anaphylaxis

A severe allergic reaction (anaphylaxis) is a life-threatening emergency. Symptoms of an allergic reaction can appear in seconds or take minutes to hours to appear.

Symptoms of severe allergic reaction may include

- Rash (hives)
- Swelling of the throat
- Confusion
- Dizziness
- Nausea and vomiting
- Abdominal cramping

If You Witness a Person Having Signs of Anaphylaxis:

- Call for emergency medical services immediately
- Check for allergy medications the person might be carrying (such as injectable epinephrine, EpiPen®). Give the drug as directed. For auto-injectable epinephrine, this consists of jabbing the auto-injector into the person's thigh and holding for about 10 seconds followed by massaging the area for 30 seconds. Antihistamine allergy medications can be given if the person can safely swallow without choking.
- If there is any vomiting, place the person on his/her side to prevent choking. (See rescue position on page 86) If the person is not vomiting and breathing comfortably, he/she can be placed on back with feet elevated.
- If the victim becomes unconscious, reassess ABCs (airway, breathing, circulation) to determine whether rescue breathing or CPR is necessary.

! WARNING: Continually monitor the victims ABCs and provide appropriate first aid until emergency services arrive.

Altitude Sickness

Travelers, hikers, and mountain climbers are at risk of altitude sickness, which is potentially harmful or fatal if ignored. Symptoms of altitude sickness can develop after ascending too rapidly to high elevation, even of less than 8000 feet (about 2500 meters).

Early symptoms of altitude sickness may include:

- Headache
- Fatigue
- Sleepiness
- Lack of coordination
- Dizziness
- Nausea and vomiting
- Symptoms of severe altitude sickness include:
- Shortness of breath
- Rapid heart rate or heart palpitations
- Coughing, sometimes with frothy sputum
- Inability to stand or sit straight up or walk in a straight line
- Bizarre or irrational behavior, such as denial of obvious symptoms

Suspected Altitude Sickness

The most important rule in managing altitude sickness is simple yet often ignored. If a person experiences symptoms of altitude sickness, he or she should not climb any higher until ALL symptoms have completely resolved. Failure to do so can cause mild altitude mountain sickness to progress to severe or fatal high-altitude pulmonary edema (fluid buildup in the lungs) or cerebral edema (swelling of the brain). Remember, it is common for persons to refuse to acknowledge symptoms of altitude sickness.

- The person should rest and keep warm
- Avoid smoking and alcohol
- Acetaminophen (Tylenol®) can be given at the usual dose for headache if the person is not allergic or intolerant to this medication
- If symptoms persist, consider descending by 1000-2000 feet (300-600 meters).
- If symptoms worsen, descend immediately and call for emergency medical services immediately
- Do not give altitude sickness medications or oxygen that is not prescribed to the person by his/her physician. This can mask the signs of worsening illness.

! WARNING: Continually monitor the victims ABCs and provide appropriate first aid until emergency services arrive.

Amputation

Amputation is the total severing of a body part from the body. This is a severe injury and always requires professional emergency medical evaluation. Call local emergency medical services immediately. A victim of amputation needs emergency medical attention without delay.

First Aid for Amputations:

- The first aid provider should use disposable gloves and other personal protective equipment
- Elevate the stump and apply direct pressure to the bleeding
- If bleeding cannot be stopped with elevation and direct pressure, place a tourniquet. Apply a tourniquet to a wound only if there is severe blood loss and death from bleeding is imminent. Incorrect use of a tourniquet can lead to loss of the limb, so tourniquets should only be used in lifesaving situations and, when possible, only by trained individuals.

Applying a tourniquet

- First aiders should use disposable gloves and other personal protective equipment
- Remember, elevation and direct pressure to stop bleeding should be attempted before use of a tourniquet.
- Use a non-elastic material such as a towel or sheet. Fold it to a width of one or two inches.
- Wrap the tourniquet around the limb, a few inches above the injury site, and tie a square knot. Use a strong item such as a stick, pipe, spoon, etc. to act as a windlass. Tie the windlass to the tourniquet with another square knot.
- Twist the knot with the windlass to tighten the tourniquet until bleeding stops.
- Secure the windlass and tourniquet structure by tying the ends to the victim's limb.
- If possible, mark the victim's forehead with a large "T" with the time the tourniquet was placed.

! **DANGER:** A tourniquet is only used on an arm or leg where there is a danger of the casualty losing his life (bleeding to death)

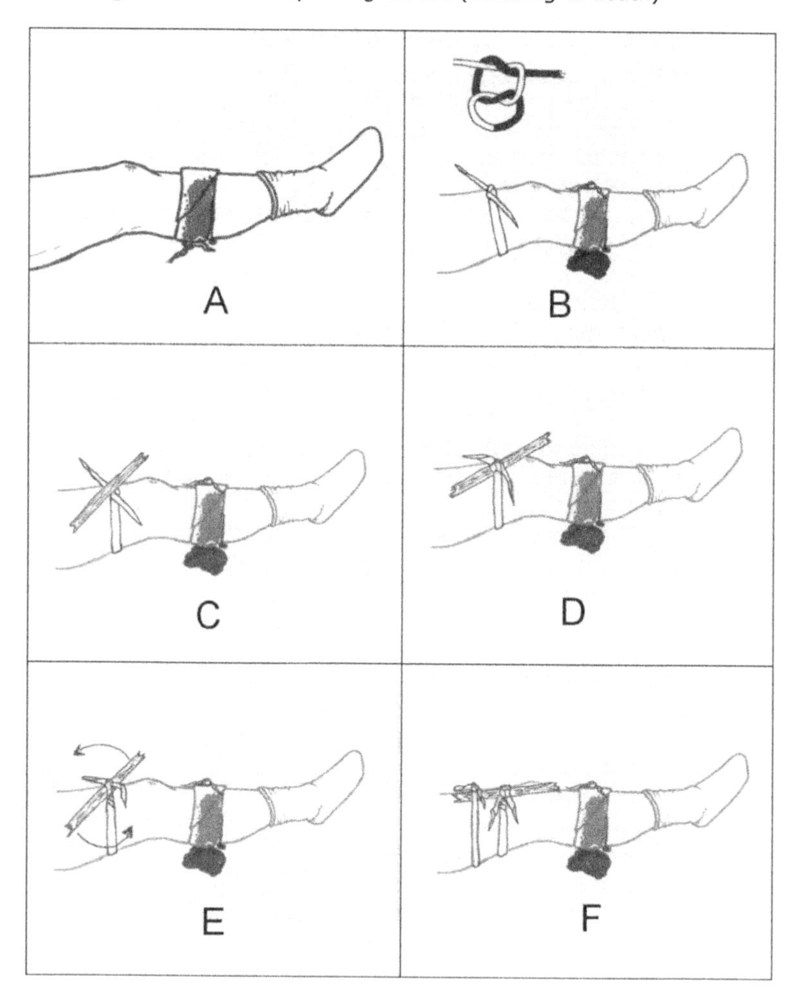

What to Do With the Amputated Limb

To save the amputated limb, wrap the part in dry, sterile gauze or clean dry cloth. Place the limb in a large plastic bag and seal the bag. Place the sealed bag into ice water. Do not allow the limb to get wet and do not place the limb directly on ice as this can damage the tissue.

Note: The amputated part should not be thoroughly cleaned and no solutions should be poured on it at this time.

! WARNING: Continually monitor the victims ABCs and provide appropriate first aid until emergency services arrive.

ASTHMA see page 103

Bites and Stings

Dog/Cat/Human/Mammals/Aquatic

Human and animal bites are common and are particularly susceptible to infection. First aid for animal bites includes ensuring safety, cleaning/dressing the wounds, and determining whether immediate physician evaluation is necessary.

- As always, first ensure safe surroundings. Move the victim (or in some cases, the biting animal) so that the area is safe.
- Always practice universal precautions: wear gloves and other personal protective equipment.
- For serious wounds, call local emergency medical services immediately
- Control bleeding by applying pressure to the area. Do not use a tourniquet unless there is massive bleeding and risk of bleeding to death (See tourniquet information in section on amputations).
- Clean less serious wounds thoroughly with soap and warm water. Both closed and open wounds should be cleaned and rinsed for several minutes.
- Dress the wounds with gauze bandages and tape. Antibiotic ointment may be applied if available.
- Over the hours and days that follow, monitor for signs of infection:
- Redness
- Drainage of pus
- Increased warmth at the area
- Increasing pain
- Swelling of the surrounding area
- Contact a physician to determine if/when the victim needs to be seen. The following types of bites should always be evaluated by a physician:
- Bite from a dog that cannot be immediately confirmed to be vaccinated against rabies
- Bites with extensive lacerations that may need stitches
- Deep puncture wounds (such as from cat bites)
- Bites involving the hands, face, or head
- Bites from wild animals such as raccoons, bats, foxes, and skunks
- Human bites that break the skin

The physician may recommend a tetanus booster shot if the wound is deep or dirty or if several years have passed since last tetanus booster.

Leeches
One recommended method of removal is using a fingernail or other flat, blunt object to break the seal of the oral sucker at the anterior end (the smaller, thinner end) of the leech, repeating with the posterior end, then flicking the leech away. As the fingernail is pushed along the person's skin against the leech, the suction of the sucker's seal is broken, at which point the leech should detach its jaws.
Internal attachments, such as nasal passage or vaginal attachments, are more likely to require medical intervention.

Treatment
After removal or detachment, the wound should be cleaned with soap and water, and bandaged. Bleeding may continue for some time, due to the leech's anti-clotting enzyme. Applying pressure can reduce bleeding, although blood loss from a single bite is not dangerous

Insect Bites and Stings
Most insect bites cause minor symptoms, but sometimes can cause significant pain and carry risk of allergic reactions and infections. Most of the time, it is impossible to determine the biting insect by the appearance of the bite.

Caring for Bites and Stings

- As always, ensure safe surroundings. Move to a safer area to avoid additional bites/stings.
- Use universal precautions. Wear gloves and other personal protective equipment when giving first aid to others.
- If a stinger is present, carefully scrape it off with a straight edge object such as a credit card or knife. Do not try and grasp and pull out a stinger (this can cause further injection of venom or infection)
- Wash the area with soap and warm water
- Apply an ice pack to the area. Never apply ice packs directly to skin; wrap ice packs in cloth. In general, cold compresses can be used for 20 minutes followed by 20 minutes off and repeated as desired.
- Hydrocortisone cream or oral antihistamine medicines (such as Benadryl) can be used as directed on package, as needed for itching. Acetaminophen or ibuprofen can be used as needed for pain.
- Watch for signs of concerning allergic reaction:
- Hives
- Swelling of the face, lips, or throat
- Confusion
- Dizziness

- Rapid heart rate
- Difficulty breathing
- If any of the above symptoms are present, seek emergency medical care immediately.

See also the section on allergic reactions for additional information

Spider Bites
Most of the world's spiders are not very dangerous to humans. Most bites can be monitored and treated as insect bites as described in the Insect Bite section above. Two spiders in the United States can cause particularly precarious problems.

Black Widow

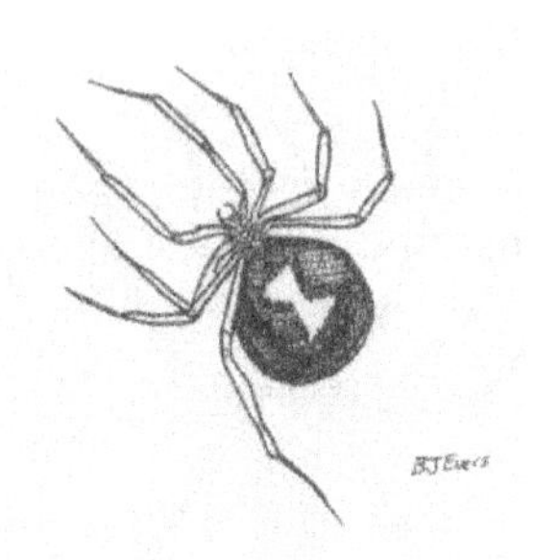

The female black widow spider is commonly found in the states in the southern half of the United States. The female is usually about 1-2 inches (2-4 centimeters) in size. It has a shiny black appearance and has a bright red hourglass shaped spot on its belly.

Black widow bites may not cause pain at the time of the bite or may feel like a thorn prick. A few hours later, however, these symptoms may develop:

- Fever and chills
- Severe abdominal pain and cramping
- Nausea and vomiting
- Sweating

Brown Recluse

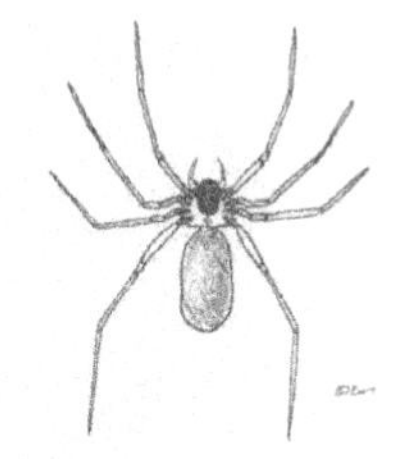

The brown recluse spider is seen in the southeastern United States. It is approximately ¼ to ¾ inches (1-2 centimeters) and tan to brown in color. It has a brown violin-shaped mark on its top side. Its bite causes a burning stinging pain followed by severe pain after six to eight hours. A blister forms and later falls off and leaves a deep ulcer. The surrounding tissue cells can die (necrosis) and heal slowly (sometimes weeks or months).
Symptoms in more severe cases include fever, chills, nausea, vomiting, body rashes, and joint pains.

Suspected Black Widow or Brown Recluse Spider Bite:

- After ensuring safe surroundings, try to make a positive identification of the biting spider. Take extreme care to avoid any further bite victims!
- Use a cold pack or ice wrapped in cloth over the site of the bite
- Seek emergency medical care immediately. Physicians will decide whether a black widow bite requires antivenin or a brown recluse bite requires treatment with steroids or other medications.
- Do not place a tourniquet!
- Using the mouth or a first aid kit suction device in attempt to remove venom is not helpful and should be avoided.

Tick Bites

There are numerous types of ticks throughout the world. Many can cause infections, some serious or even fatal. In the United States, certain ticks can cause Lyme disease or Rocky Mountain Spotted Fever. Overall, the chance of getting a serious infection after a tick bite is small. Safe appropriate removal of the tick and monitoring for signs of infection, however, is essential.

How to Remove the Tick:

- First aiders should wear gloves if available.
- Use tweezers to grasp the tick as close to the skin surface as possible. Be careful not to crush the tick's head.
- Do not squeeze, crush, or burn the body of the tick. This can cause injection of infectious fluids into the person's skin.
- Pull the tick away from the skin gently but firmly. Do not make any twisting or jerking motions with the tweezers.
- If possible, save the tick in a jar or make note of its size and color. This could help doctors later if symptoms of illness develop.
- After tick removal, wash the area thoroughly with soap and water. The person that removed the tick should also wash hands well.
- If any parts of the tick's head or mouth are embedded in the skin, leave them alone, they will be extruded on their own.

Following a tick bite and tick removal, it is important to monitor for signs of illness and infection. See a doctor immediately if any of the following symptoms develop:

- Fever
- Rash
- Muscle or joint aches
- Stiff neck
- Flu-like symptoms
-

Snakebites

Most snakebites are not fatal; however several species of snakes throughout the world have venom that is harmful to humans. Since few people have expertise in snake identification, we recommend all snakebite victims be evaluated by medical experts. There are important steps to take as prior to arrival of emergency medical services.

- As always, first ensure safe surroundings. Get the victim and first aiders away from the snake. Wear gloves and other personal protective equipment.
- Call for emergency medical services immediately.
- Try to keep the victim and people around him calm.
- Do not attempt to capture or get close to the snake! Such attempts have led to a second or even a third person being bitten. Even a dead snake has a bite reflex that can last for an hour or more after its death. Note the snake's colors and markings or photograph the snake from a safe distance.
- Immobilize the bitten arm or leg. Do not elevate the limb. Keep the site of the bite below the level of the victim's heart.
- Remove tight clothing and jewelry because swelling can occur rapidly.
- Do not use a tourniquet or apply ice!
- Do not cut open the wound or attempt to suck venom from the wound!
- If the victim cannot get to medical care within 30 minutes, apply a bandage a few inches above the bite location to slow circulation of venom. The bandage should not cut off blood circulation to the limb. It should be loose enough that a finger can slip between the bandage and the skin.

Anatomic features of vipers

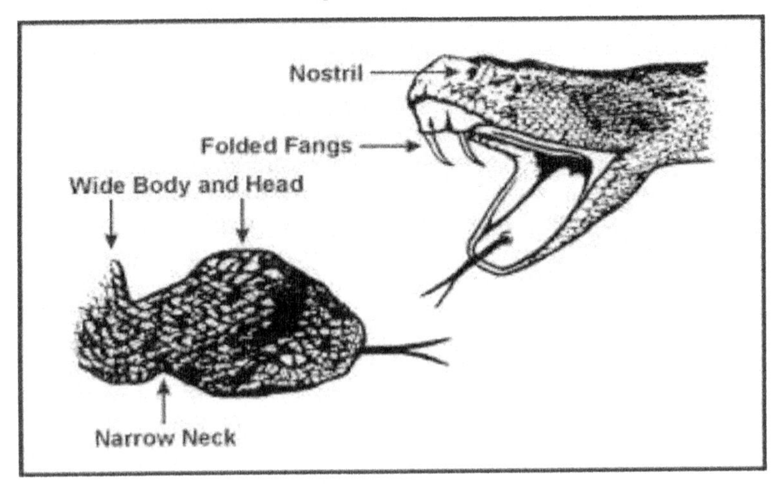

Features of Cobras, Kraits, and Coral Snakes

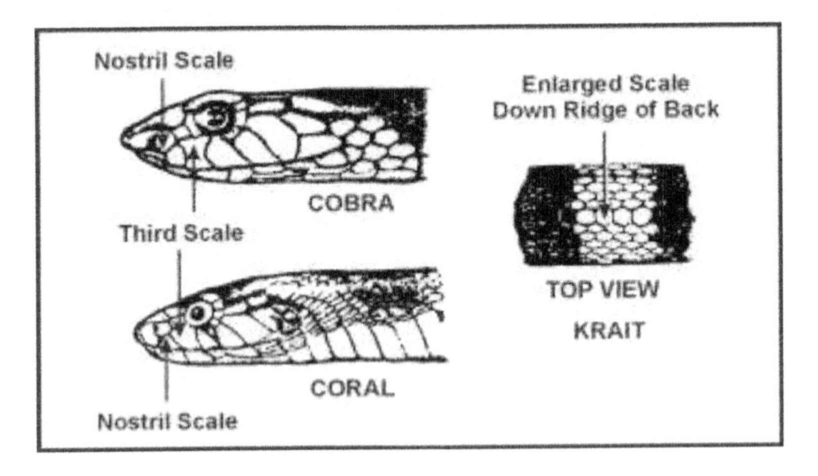

Jellyfish Stings

There are hundreds or thousands of jellyfish species in oceans around the world. Stings from jellyfish may cause minimal symptoms or cause severe pain or severe, potentially fatal allergic reactions.

- As always, first ensure safe surroundings. First aiders and victims should be safely away from the water. Gloves and personal protective equipment should be used.
- Continually monitor for symptoms of severe allergic reaction. (See section on allergic reactions for details).
- If at any point there is doubt about the victim's status or symptoms are worsening, contact emergency services immediately. If the jellyfish is known or suspected to be

a box jellyfish (a particularly dangerous type of jellyfish) call for emergency services without delay.

- First, rinse the skin with vinegar or seawater. Next, the tentacles should be removed. Stinging continues as long as contact with skin remains. Use a gloved hand, stick, or tweezers to carefully remove tentacles. Do not touch the tentacles with bare hands! For large amounts of tentacles, allow physician/medical professionals to perform removal.
- If possible, soak the area in hot water (as hot as tolerable; take extreme care not to burn or scald the skin).
- Seek professional medical evaluation for further treatment as soon as possible.
- Additional notes:
- Portuguese Man of War and Bluebottle are not technically jellyfish, but first aid for stings is as above. Medical evaluation should be sought immediately.
- Contrary to popular belief, urinating on stings does not seem to be an effective treatment.
- For persistent pain and stinging, over-the-counter pain relievers may be used as directed on the packaging. Some experts recommend applying a baking soda-water paste to the sting sites.
- Over the hours and days that follow the sting, monitor for signs and symptoms of infection.

See Animal Bites, page 92 , for a list of symptoms

Other Marine Related Injuries

Related injuries include abrasions, stings, puncture wounds, etc. from other marine organisms such as anemones, urchins, and corals. These injuries can be treated like other land-based injuries:

- Clean areas gently with soap and water
- Monitor closely for signs and symptoms of allergic reaction or infection (See above).
- Over the counter pain or allergy medication can be used for mild symptoms. Antibiotic ointment can be applied if desired.
- Seek emergency medical care for any concerning symptoms or if victim's status is worsening or uncertain.

! WARNING: Continually monitor the victims ABCs and provide appropriate first aid until emergency services arrive.

Blast Injury

Explosions from bombs or other sources can produce multiple types of life-threatening injuries on many people at one time. Victims of explosions who survive the initial blast may still be in danger from several types of injury. The intense highly-pressurized impulse from a blast can cause injury by rapid pressure changes, especially in gas-filled body structures (lungs, digestive tract, middle ear, for example).

"Blast lung" is the most common fatal injury of blast injury victims who survive the initial explosion. Chest pain, shortness of breath, cough, and coughing up blood are some symptoms of blast lung. Onset of symptoms is sometimes delayed for up to 48 hours. Blast abdominal injury may be present in any victim with abdominal pain, nausea and vomiting, testicular/groin pain, or rectal pain. Any victim of a blast should be evaluated in an emergency department immediately.

Any number of injuries can be caused by explosion. For information on first aid for specific types of injury refer to the additional sections in this book:

- Penetrating injuries
- Concussion
- Eye injuries
- Fractures
- Burns
- Anxiety/Hyperventilation
- Amputations
- Breathing emergencies
- Cuts and Bleeding

! WARNING: Continually monitor the victims ABCs and provide appropriate first aid until emergency services arrive.

Bleeding

The longer a person bleeds from a major wound, the less likely he will be able to survive his injuries. It is, therefore, important that the first aid provider promptly stop the external bleeding. In evaluating the casualty for location, type, and size of the wound or injury, cut or tear his clothing and carefully expose the entire area of the wound. This procedure is necessary to properly visualize injury and avoid further contamination.

Clothing stuck to the wound should be left in place to avoid further injury. DO NOT touch the wound; keep it as clean as possible.

Before applying a dressing, carefully examine the victim to determine if there is more than one wound. (Gunshot, Stabbing and Bombing Victims in Particular) Hold the dressing directly over the wound.

If bleeding continues after applying a dressing direct manual pressure may be used to help control bleeding. Apply such pressure by placing a hand on the dressing and exerting firm pressure for 5to 10 minutes .Elevate an injured limb above the

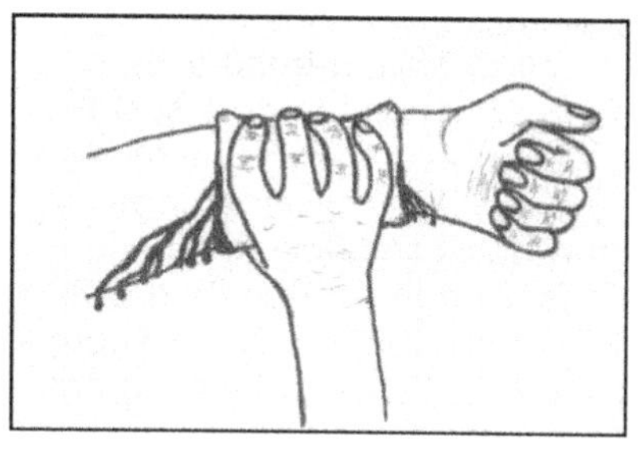

level of the heart to reduce the bleeding

! WARNING: Do not elevate a suspected fractured limb unless it has been properly splinted.

If the bleeding stops, check shock; administer first aid for shock as necessary. If the bleeding continues, apply a pressure dressing.

Pressure Dressing

Pressure dressings aid in blood clotting and compress the open blood vessel. If bleeding continues after the application of a field dressing, continue manual pressure, and elevation, then a pressure dressing must be applied as follows:

- Place a wad of padding on top of the field dressing, directly over the wound
- Keep the injured extremity elevated. Improvised bandages may be made from strips of cloth. These strips may be made from T-shirts, socks, or other garments
- Place an improvised dressing (or cravat, if available) over the wad of padding. Wrap the ends tightly around the injured limb, covering the previously placed dressing
- Tie the ends together in a nonslip knot, directly over the wound site Do Not tie so tightly that it has a tourniquet-like effect.
- If bleeding continues and all other measures have failed, or if the limb is severed, then apply a tourniquet.
- Use the tourniquet as a **last resort**. When the bleeding stops, check for shock; administer first aid for shock as necessary.
- Distal end of wounded extremities (fingers and toes) should be checked periodically for adequate circulation. The dressing must be loosened if the extremity becomes cool, blue, or numb.
- If bleeding continues and all other measures have failed(dressings and covering wound, applying direct manual pressure, elevating the limb above the heart level,

and applying a pressure dressing while maintaining limb elevation) then apply digital pressure.

See Tourniquet under Amputation, page 90.

Digital Pressure

Digital pressure (often called "pressure points") is an alternative method to control bleeding. This method uses pressure from the fingers, thumbs, or hands to press at the site or point where a main artery supplying the wounded area lies near the skin surface or over bone (Figure below). This pressure may help shut off or slow down the flow of blood from the heart to the wound and is used in combination with direct pressure and elevation. It may help in instances where bleeding is not easily controlled, where pressure dressing has not yet been applied, or where pressure dressings are not readily available. *WARNING: Continually monitor the victims ABCs and provide appropriate first aid until emergency services arrive.*

Pressure Points

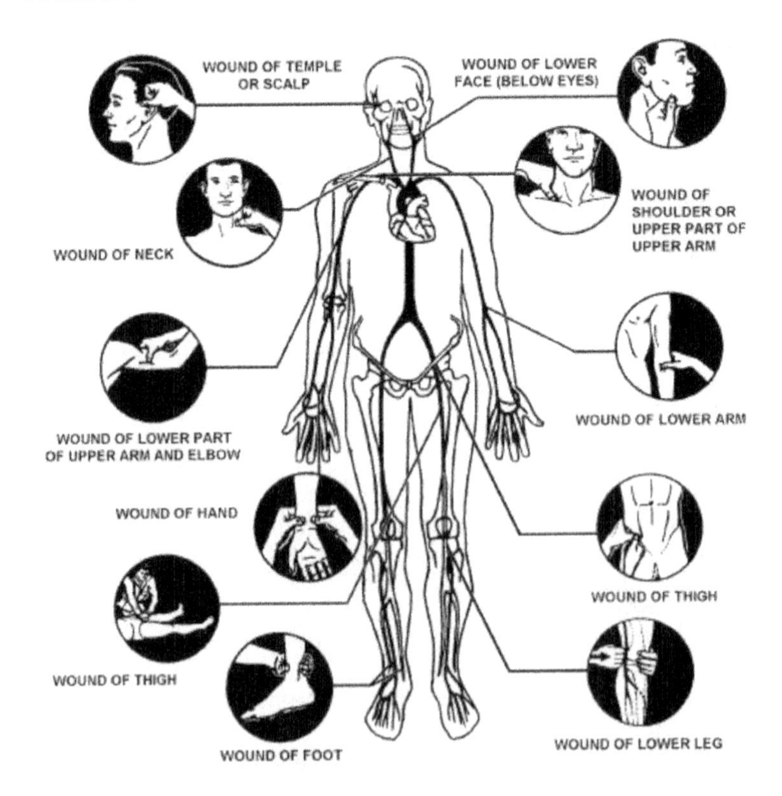

Breathing Emergencies

The body requires effective breathing to maintain survival. Any compromise of normal breathing can become a life-threatening emergency.

There are numerous causes of respiratory distress including asthma, allergies, choking, trauma, heart disease, brain injury, and shock. There are common signs and symptoms of respiratory distress, including:

- Abnormally fast or slow breathing
- Gasping for air, inability to catch breath
- Abnormal breathing sounds such as grunting or wheezing
- Confusion, dizziness, anxiety
- Pale or blue skin, most often starting in the fingers or around the lips

Persons with respiratory distress should be kept calm, assisted with taking their medications, and comfortably positioned in a location with adequate air ventilation. Expert medical care should be sought immediately.

Asthma Attack

Sometimes asthma attacks can be treated with inhaler medications, efficient breathing, and relaxation. Other times, treatment requires emergency transport to the hospital.

Treating Asthma Attacks:

- Try to keep the person calm. Increased muscle tension and anxiety can make breathing less effective.
- Have the person start pursed-lip breathing—exhaling through pursed lips (as if blowing out a candle). Breathing should be slow and not forceful.
- Assist the person with using their rescue inhaler. These are usually beta-agonist medications such as albuterol (Brand names include Proventil, ProAir, Ventolin, others) or levalbuterol (Brand name Xopenex) and others. In general, two puffs of these medications can be used up to every twenty minutes as needed.
- Separate the person from asthma triggers if known. Persons with asthma often know what can trigger their attacks. Some examples of asthma triggers include dusts, pet dander, perfumes, smoke, exercise, and cold air.
- Assess response to treatment. Evidence of good response may include improvement of work of breathing, ability to speak in full sentences, and decreasing wheezing and coughing (caution: decreasing wheezing can sometimes be a sign of worsening asthma attack!). Signs of poor response or worsening attack can include more rapid breathing, more difficulty with breathing, difficulty

speaking, and blue color in lips or fingers.

- When in doubt or if response is poor, immediately seek emergency medical care.

Choking

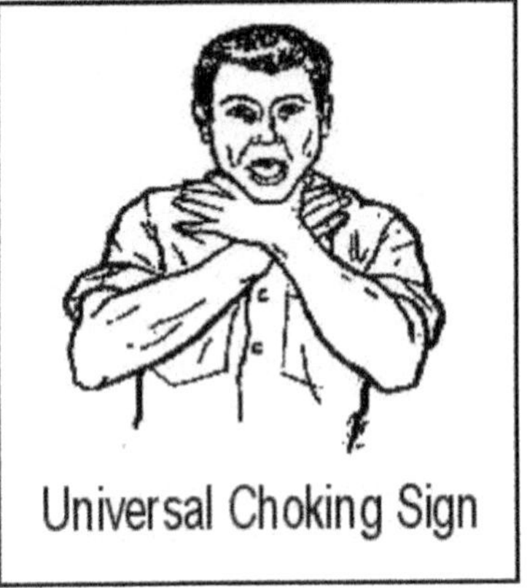
Universal Choking Sign

The "universal choking sign" is the choking victim clutching his neck with his hands. Other signs of choking include: difficulty or inability to breathe, cough, or speak.
If the person is able to cough, encourage him to continue coughing. If the choking victim requires further assistance:

- Tell the victim to lean forward and using the heel of the hand, give five firm blows to the back
- Next, give five fast thrusts of the abdomen by reaching around the person and placing a fist just above the level of the belly button. Grasp the fist with the other hand.
- Repeat until the victim begins coughing or breathing or the object being choked on is forced out of the airway.

Respiratory failure is a medical emergency. Signs of respiratory failure include loss of consciousness, loss of breath sounds and chest movements, and blue color of the skin. For respiratory failure begin rescue breathing and CPR immediately.
See sections on CPR, page 107, for additional information.

Inhalation Injury
Inhalation injury can be caused by a large variety of inhalants including smoke and other gases or fumes.

- Assess the scene. Make sure you can give first aid safely without placing yourself in danger. Safely move the victim away from the offending inhalant to an open, well-ventilated space.
- Call for emergency services.
- Assess the ABC's. Make sure the victim is breathing effectively and has fresh air.
- Try to make the victim's breathing as comfortable as possible. Change his positioning to the recovery position (see page 86) and loosen clothing, making sure to keep the victim warm.
- Reassess ABCs frequently. If the person has compromised airway, breathing, and/or circulation, begin CPR as

described on page 107.
- Continue monitoring and supportive care until medical help arrives.

Hyperventilation

Rapid shallow breathing should never be assumed to be explained by "hyperventilation" (due to panic attack, stress, or anxiety) unless the victim has been diagnosed with a hyperventilation syndrome by their doctor. Any unexplained breathing difficulty should be evaluated by emergency medical personnel immediately.
First aid for hyperventilating person with known diagnosis of a hyperventilation syndrome: The goal is to calm the person and encourage and coach the person toward a more comfortable breathing pattern.

- Encourage slow deep breaths.
- Ask the person to breathe with you and try to follow your breathing
- If the person does not calm, seek medical help.
- Breathing into a paper bag is not recommended.

! WARNING: Continually monitor the victims ABCs and provide appropriate first aid until emergency services arrive.

Burns

First aid for burns varies based on the severity of the burn. In general, minor burns are small, superficial, and not located in high-risk areas (listed below). A burn should be considered a major burn if it is:

- Deep (if muscle, fat, or bone is visible or if there is charring of tissue)
- Larger than 3 inches in diameter
- Painless
- Located on face/head, groin, over a joint, or covers a considerable area of hands or feet.

Minor Burns:

- Pour cool water over the burn for 10 minutes, or place cool compress (towel soaked in cool water) on the burn wound. Do Not Apply Ice To A Burn Injury.
- Dress the burn loosely with sterile gauze. If blisters are present, do not break them open.
- Over-the-counter pain medications should be sufficient for pain management. Monitor for signs of infection, (worsening redness or pain, fever, pus or fluid drainage).
- If signs of infection are present or if there is any question of a possibly more severe or worsening injury, professional

medical attention should be sought.

Major Burns:
- Remember, always survey the scene and ensure safe surroundings first!
- Call for emergency medical services immediately.
- Cover the area with cool wet towels or gauze.
- Do not remove the victim's clothing.
- Do not immerse large burns in water.
- Do not apply ice to a burn injury. Help the victim to remain calm until help arrives.
- Assess for possible lung injuries from inhalation of smoke or other toxins.
- First aid for Chemical burns -
- Flush the burned area with cool water for 20-30 minutes.
- Cover the area with a cool wet towel or dress loosely with sterile gauze.
- Call poison center 1-800-222-1222 if not sure of the toxicity of the chemical.
- Seek emergency care for a major burn.

Sunburn

Symptoms usually develop several hours after sun exposure and may include;
- Redness and swelling of skin
- Pain
- Warmth of skin
- Blisters

If large areas of skin are affected by sunburn, additional possible symptoms include:
- Fever
- Headache
- Fatigue

Sunburn Treatment
- Take a cool shower or bath. One-half to one cup (120-240ml) of cornstarch, oatmeal, or baking soda may be added to the bathwater and may provide additional relief
- For pain, you may take acetaminophen (Tylenol) or ibuprofen (Motrin, Advil) in the usual over-the-counter doses, if necessary.
- Drink plenty of water or electrolyte-containing sports drinks.
- Aloe vera lotion or gel can be applied to sunburned skin several times per day, if desired.
- If blisters are present, leave them intact. If blisters break on their own, they may be covered with antibiotic ointment.
- If symptoms worsen or if there are signs of infection, seek medical evaluation immediately.

! WARNING: Continually monitor the victims ABCs and provide appropriate first aid until emergency services arrive.

Cardiopulmonary Resuscitation (CPR)

Adult CPR: Quick Review

- Is the victim unresponsive, not breathing, or gasping?
- Call 911 or Emergency Medical Services
- Begin chest compressions
- Push Hard, Push Fast
- If not trained in rescue breathing, continue hard fast chest compressions until AED arrives or EMS takes over.

If Rescuer Has Emergency CPR Training

- Initiate breathing after 30 compressions
- After 30 compressions, pause for 2 rescue breaths
- Continue 30 compressions : 2 breath cycles, pausing every 2 minutes to evaluate for breathing/movement or to allow the AED to evaluate for rhythm.
- Continue until victim is breathing/moving or until emergency medical team arrives

Detailed CPR Review

CPR is a technique that can be used in emergency medical situations when a person's heartbeat has stopped. The most recent recommendations by the American Heart Association are summarized below. We recommend taking a formal accredited CPR training course. Just about anyone can save a life with CPR, with or without advanced training.

It is better to do something than nothing! Data shows that an attempt at CPR (even by untrained persons) is better than no attempt at all.

Automated external defibrillators (AED) are medical devices that can deliver electrical shocks to the heart to attempt to reset a heart's normal heartbeat and rhythm. These devices are becoming more and more readily available in public places. Many CPR courses have sections on using AEDs.

The following instructions are for performing CPR on an adult. (CPR for children and infants is discussed at the end of this section.)

To Determine Whether CPR Is Necessary:

- Determine whether the person is conscious. If she/he appears unconscious, tap the person on the shoulder and ask loudly "Are you OK?"
- If the person is unresponsive ask someone to call for

emergency services immediately. If you are alone with the victim and have immediate access to a telephone, call for emergency medical services before starting CPR*.
- If the victim is unconscious because of drowning or suffocation, perform CPR for one minute before calling emergency medical services.
- If an AED and trained user are immediately available, attach the AED to the victim. Deliver one shock if instructed by the AED before starting CPR.

The ABC's (Airway, Breathing, Circulation)
- Quickly assess airway, breathing, and circulation:
- Keep victim's head and neck aligned with body

To open the airway, place the victim on his back with the head tilted back. Gently push the chin forward. This allows air to flow to and from the lungs through both nose and mouth. If the victim has traumatic injuries, use extreme care and minimize movement of the neck.
- If the victim is breathing, place him in the recovery position (see page 86).
- If you have had appropriate training, check for pulse.
- If the victim is not breathing or is gasping, begin chest compressions.

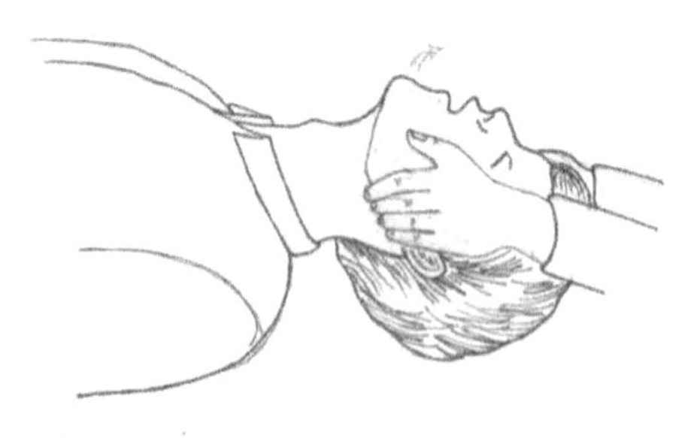

Chest Compressions: Push Hard, Push Fast

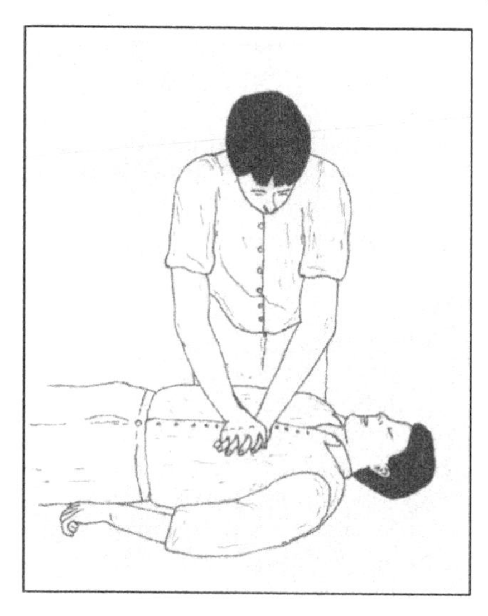

- Remember, ask for and listen closely for instructions from the EMS dispatcher when available.
- Place the heel of one hand on the center of the victim's chest. Put the other hand on top of the first and interlock the fingers. Position the shoulders directly above your hands
- Compress the chest about 2 inches. Push Hard and Push Fast. The compression rate should be 100 per minute—or about 2 compressions per second. Count compressions aloud as you go.
- Allow the chest to fully recoil after each compression. Keep hands on the chest during compressions; do not let your hands "bounce"
- You may feel popping or crunching sounds when you start. Keep going! If the victim has cardiac arrest, CPR is more important than avoiding broken ribs.
- If trained first aiders are present, stop after 30 compressions and give 2 rescue breaths. This is one cycle of CPR.
- If trained responders are not available, continue Hard Fast chest compressions with minimal interruptions until additional help arrives.
- Continue cycles of CPR for about 2 minutes, then stop and check for breathing or movements. If an AED is available, follow its instructions. It may deliver a shock.
- Resume 2 minute cycles of CPR until there are signs of movement or breathing or until emergency medical services arrives
- Performing chest compressions is hard work! After a few cycles allow another first aider to take over to allow you to rest.

Rescue Breathing

- If you have not had training in rescue breathing, continue Hard Fast chest compressions!
- If you have been trained in rescue breathing, position the

airway as described above. Make a good seal on the airway with your mouth or rescue breathing device. Deliver a one second breath while watching for chest rise. If the chest rises, deliver a second rescue breath. If there is no chest rise, reposition the airway before delivering a second breath.

- Continue 30 chest compressions to 2 rescue breath cycles until there are signs of movements or until EMS arrives .

! WARNING: Continually monitor the victims ABCs and provide appropriate first aid until emergency services arrive.

CPR For Children and Infants

Call 911 or Emergency Medical Services

CPR for children age 1 thru 10 is almost the same as for adults. There are some important differences:

- If no other help is present, perform 5 cycles of CPR (chest compressions and breaths; this should take about 2 minutes) before pausing to call emergency services or attaching an AED
- Use less forceful breaths with rescue breathing
- Perform chest compressions with only one hand
- After five cycles of CPR (as in an adult, use the 30 compression to 2 breath cycle) place the AED pads if available. Use pediatric pads if available (if not, use adult pads). Note: Rescuer teams with advanced training may use a 15 to 2 cycle.
- Continue CPR until emergency medical professionals arrive or until the child shows signs of breathing and circulation (breathing, moving, pulse).

Performing CPR on an Infant

Call 911 or Emergency Medical Services

Respiratory and/or cardiac arrest in otherwise healthy infants often results from airway obstruction (often related to choking or drowning). A thorough evaluation of the ABCs is critical.

Airway and Breathing:

With the infant lying on his back with his head tipped back, look, listen and feel for evidence of breathing (as described above). If he is not breathing, begin rescue breathing immediately.

Rescue breathing for a baby:

Seal your mouth over the infant's nose and mouth. Deliver a one-second breath using only the air from the mouth. Use cheeks to deliver the breath. Do not force air into the infant using your lungs. As the breath is delivered watch for the chest to rise as the lungs fill with air. If no chest rise is seen, reposition the infant using the head tilt and chin lift maneuvers, then resume rescue breaths. If there is still no chest rise seen with breaths, examine the mouth and throat for evidence of foreign object. Sweep the mouth with a finger and remove foreign material if possible.

Performing chest compressions on a baby:
Place two fingers on the center of the infant's chest at the level of the nipples. Compress one-third to one-half the depth of the chest. Perform compressions at the same 100-per-minute rate as in an adult. Count compressions aloud. After every 30 compressions, perform 2 rescue breaths.
Continue CPR until emergency medical personnel arrive or until the baby shows signs of breathing and circulation.

! WARNING: Continually monitor the victims ABCs and provide appropriate first aid until emergency services arrive.

Chest Trauma

See also Breathing Emergencies (page 103), Puncture Wounds (page 100), CPR (page 107). Crush injury (page 113)
Blunt trauma to the chest can result in various injuries including bruising of the chest wall, ribs, or internal organs. Fracture of one or more ribs can cause severe pain and may affect breathing. Severe internal injury may be present even if the external chest appears normal.
If any of the following are present, seek emergency medical attention immediately:

- Difficulty breathing
- Coughing up blood
- Severe pain
- Dizziness, confusion, lightheadedness
- Deformity of the chest
- Asymmetric or abnormal chest movements with breathing

Chest injury

- Keep the injured person calm and still
- Assess the ABCs (See section on the ABCs of first aid)
- Do not move the victim's neck
- Gentle support with a hand or pillow over the injured area may ease breathing. Do not use a binder or elastic bandage that wraps around the entire chest as this can make deep

breathing difficult.

- Gentle repositioning may make breathing more comfortable for the injured person. Positioning with the injured side down may ease breathing.
- Reassess ABCs frequently. Any breathing difficulty should be considered a medical emergency. Stay with the victim until help arrives.

! WARNING: Continually monitor the victims ABCs and provide appropriate first aid until emergency services arrive.

Penetrating Chest Wounds

Penetrating wounds like stab wounds or gunshot wounds can cause severe internal injuries even with a very small external entry wound. Spinal cord injuries should always be considered and victim should not be moved unless absolutely necessary. Positioning/repositioning of victim should be done with extreme caution.

Treating a Penetrating Chest Wound

- Assess scene safety.
- Ensure safe surroundings before starting first aid.
- Emergency medical services should be called immediately.
- First aiders should wear disposable gloves and other personal protective equipment.
- Assess and treat ABCs before treating other injuries.
- Pressure should be placed on the penetrating chest injury site by a first aider or by the victim. This helps to slow bleeding and helps prevent air from entering the chest cavity.
- The victim should be placed in a relaxed slanted upright position with the wounded side slanting downward. Help the victim to try to breathe calmly.

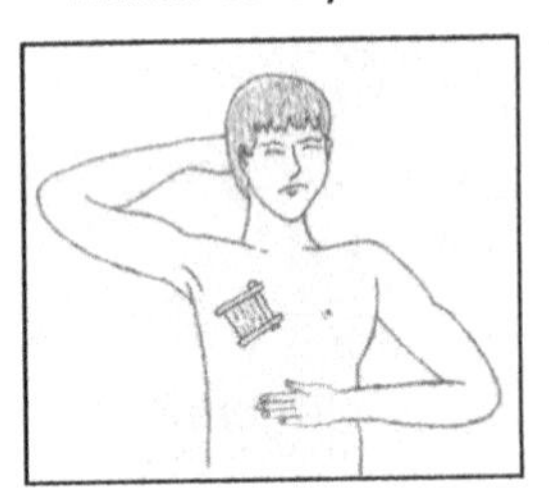

- If first aid supplies are available, a square or rectangle shaped bandage can be placed over the wound. The bandage should be sealed (taped) on three sides. The fourth side can be left unsealed (See figure). It may assist to gently lift the unsealed side while the victim is exhaling. This will allow air to escape the chest cavity.
- When applying additional bandages, cover the soaked dressings with new bandages (do not remove the old soaked bandages).
- Reassess ABCs frequently.
- Start rescue breathing and/or CPR as necessary.

! WARNING: Continually monitor the victims ABCs and provide appropriate first aid until emergency services arrive.

Chest Pain

See Heart Attack, Page 124

Choking

See Breathing Emergencies on Page 104.

Concussion/ Head Injury

An impact to the head can result in a wide range of injuries. Serious brain injuries including bleeding within the head (intracranial bleeding) can be difficult to diagnose without a physician's evaluation in a hospital. First aiders of a victim of head trauma should monitor closely for several signs and symptoms that could indicate a severe brain injury.
For a person who has sustained an impact to the head, emergency medical services should be called immediately if any of the following symptoms are present:

- Loss of consciousness
- Persistent confusion
- Loss of balance
- Slurred speech
- Severe headache
- Unequal pupil size
- Black/blue discoloration around the eyes or behind the ears
- Bleeding or fluid draining from ears
- Repeated vomiting
- Seizure

Treating Head Injuries

- Keep the victim calm and still
- Assess the ABCs (See section on the ABCs of first aid)
- Have the victim lie down on his back, with head and shoulders slightly elevated
- Avoid moving the neck
- Stop bleeding to lacerations by applying pressure to the area. Do not apply pressure if skull fractures are suspected.
- Continually monitor for alteration of consciousness and breathing. Remember the ABCs of first aid and begin CPR if necessary (See sections on ABCs and CPR). Remain with the victim until help arrives.

! WARNING: Continually monitor the victims ABCs and provide appropriate first aid until emergency services arrive.

Crush injury Crush Syndrome Compartment Syndrome

A Crush Injury happens when a body part is subjected to a high degree of force or pressure, and can occur from falling debris, industrial equipment, vehicles and other means. Typically affected areas of the body include lower extremities, upper extremities, and trunk.

A Crush injury can cause the following damage and more:
Bleeding
Bruising
Fractures
Lacerations
Crush syndrome
Compartment Syndrome

Crush syndrome is localized crush injury with systemic manifestations. These systemic effects are caused by a traumatic muscle breakdown and the release of potentially toxic muscle cell components and electrolytes into the circulatory system. Crush syndrome can cause local tissue injury, **and can result in kidney, heart and other problems**.

First Aid
Make sure area is safe!

! WARNING: Continually monitor the victims ABCs and provide appropriate first aid until emergency services arrive.

Stop bleeding by applying direct pressure.
Do survey for other injuries such as neck and spine injuries and treat accordingly.
Cover the crush area with a wet cloth or bandage, then raise the area above the level of the heart, if possible.
Keep victim warm
Seek immediate medical attention

Compartment syndrome is caused by increased pressure in an arm, leg or hand that causes serious tissue, nerve, blood vessel, and muscle damage. There is severe pain when you raise the affected area, and pain medicine does not help. In more severe cases, symptoms may include:

Swelling
Numbness and tingling
Decreased sensation
Weakness
Paleness of skin
Severe pain that gets worse

Seek immediate medical attention

Cuts, Scrapes, and Bleeding

Minor scrapes and cuts can often be treated without a trip to the clinic or emergency room. Good first aid is essential, however, to avoid complications that can make a minor injury into a more serious situation.

Minor Cuts and Scrapes:

- Ensure safe surroundings. Use gloves and other personal protective equipment.
- Stop any bleeding by applying pressure to the area for several minutes. Avoid the urge to frequently check to see if bleeding has stopped. This can damage the fresh clot that is forming. If there is a large amount of bleeding, spurting blood, or if bleeding cannot be stopped within 20-30 minutes, seek medical attention immediately.
- Clean the area with warm water. A small amount of soap may be used but may cause irritation to the wound. Gently remove any surface debris with tweezers or by gently wiping with a clean cloth. Wounds with large embedded debris or very dirty wounds should be evaluated by a physician. Cleaning minor wounds with alcohol, hydrogen peroxide, or iodine is not necessary.
- If desired, apply a layer of antibiotic ointment over the wound.
- Dress the wound with bandages. Band-Aid or similar bandages work well for small wounds. Larger wounds can be dressed using gauze pads and tape. Change the dressing at least once a day. Change the dressings if they become dirty or wet.
- Every time dressings are changed, look closely at wounds for any signs of infection:
- Increasing redness
- Swelling
- Drainage of pus from the wound
- Worsening pain
- Warmth at the area
- Contact a physician if any of these symptoms develop.
- Tetanus booster shots are recommended for all adults every 10 years. If it has been more than 5 years since last tetanus booster or if wounds are deep or dirty, a tetanus booster may be necessary. Tetanus booster injections should be completed within 2 days (48 hours) of injury.

Dental Injuries

Toothache

Most toothaches are related to dental caries (cavities) caused by tooth decay. This decay is related to the carbohydrates in the diet and the way sugars and starches are metabolized by bacteria in the mouth. When tooth enamel is worn away by chemicals produced by bacteria, cavities are created. Often the first symptom a person has is pain, commonly elicited by cold or hot liquids.

Minor toothaches can be treated using the recommendations below. At some point, however, the toothache will need the evaluation and treatment of a dentist.

Minor Toothaches:

- Rinse mouth with mouthwash or warm water
- Take over-the-counter pain medicine such as acetaminophen or ibuprofen as directed on the packaging
- Use a toothbrush and dental floss to gently remove food fragments from between the teeth
- Over-the-counter topical anesthetic medicine (such as benzocaine containing gels) can be applied to the area of pain where the tooth meets the gums

If any of the following symptoms are present, seek dental evaluation immediately:

- Signs of infection: severe pain (especially with biting), drainage of pus or sour tasting fluid, swelling of gums
- Difficulty swallowing
- Pain persists for several days

Tooth Loss

After a tooth is knocked out, immediate dental care is essential. Dentists can sometimes replace the tooth if appropriate steps are taken immediately after the tooth loss.

After a tooth has been knocked out:

- Immediately plan for and seek care at a dentist's office or emergency room.
- Hold the tooth by the top (crown) only. Do not touch the root.
- Do not scrape any debris from the tooth
- Gently rinse the tooth in a bowl of warm water. Do not hold the tooth under running water
- Try to replace the tooth into its socket. If it does not go all the way in to the socket, gently bite down to push the tooth into place. Hold the tooth in place until the dentist's exam.
- Sometimes replacing the tooth is not possible. If this is the case, transport the tooth in a container of milk, saliva, or warm saltwater (1/4 teaspoon of salt in 16 ounces of

water)

Diabetic Emergencies

Diabetes mellitus is a medical condition in which blood sugar (glucose) concentrations can be abnormally high due to abnormality in the body's normal glucose regulation. Insulin is a hormone that functions to lower blood sugar levels.
To manage blood sugar, persons with diabetes may be on oral medications, injectable medications, both, or no medications at all. Urgent complications of diabetes include abnormally high blood sugar (hyperglycemia) and abnormally low blood sugar (hypoglycemia).
Hypoglycemia can be caused by a higher than required dose of insulin or oral diabetes medicine (which can drop the blood sugar to dangerously low levels). Low blood sugar can also develop if a person taking blood sugar lowering medicine(s) misses necessary nutritional intake that balances effects of the medication.
Hyperglycemia in a diabetic person may be caused by missed doses of diabetes medications, effects of other illnesses, or side effects of other medications
Symptoms of each are similar and may include:

- Confusion, irritability, and/or lightheadedness
- Sweating
- Cool clammy skin
- Rapid shallow breathing

Diabetic Treatment

- If the person is unresponsive, call for emergency medical services immediately.
- If unconscious, place them into the recovery position (see page86) and ABC's should be monitored until help arrives. If he is conscious and alert, ask what is wrong. He may be able to tell you what needs to be done to help.
- Many first aid kits contain glucose tablets or gel. This can be given if the person can safely swallow. If first aid glucose is unavailable, a small amount of juice or candy can be given.

! WARNING: Continually monitor the victims ABCs and provide appropriate first aid until emergency services arrive.

Diarrhea, Nausea, and Vomiting

There are several causes of gastroenteritis (inflammatory illness of the stomach, intestines, and colon). Causes include:

- Infection (viruses, bacteria, parasites)
- Toxins (usually produced by bacteria) from contaminated

food or water
- Side effects of medication
- Food intolerances

In addition to nausea, vomiting, and diarrhea, additional symptoms of gastroenteritis may include abdominal cramping, bloating, and fever.

Treating Gastroenteritis:
- Drink plenty of fluids. Water or sports drinks may be best. Do not drink large amounts of fluids in a short amount of time. This prevents the stomach from becoming full, which can make nausea worse.
- Limit food intake on the first day. With gastroenteritis, fluids are more important than food. Allow the GI tract to rest for the initial part of the illness.
- Return to regular foods slowly. Start with bland, soft foods and advance back to normal meals slowly.
- Rest. Allow your body time and rest for recovery.

Often, symptoms with improve and resolve over 2-3 days. If any of the following symptoms develop, seek medical attention immediately:
- Inability to drink and keep down fluids for more than several hours.
- Fever above 101 F (38.3 C)
- Vomiting or diarrhea that persists for more than 3 days
- Lightheadedness or dizziness
- Severe abdominal pain
- Bloody diarrhea or vomit

Drowning (Near-Drowning)

Near drowning is always a medical emergency. It often occurs to victims without necessary swimming skills or when a victim falls through ice.

Near Drowning:
- As always, first ensure safe surroundings. Make sure any potential rescuers are strong enough and skilled enough to enter water to rescue a victim. Never walk onto ice to try to rescue a person who had fallen through ice. Do not attempt rescue unless you are sure it can be done safely.
- Call for emergency services immediately!
- If the victim is not breathing, begin rescue breathing immediately (for properly trained individuals, rescue breathing can be done while victim is still in the water). Give rescue breaths every 5 seconds.
- Use extreme caution when moving a victim of near drowning. Assume the victim could have a neck or spine injury.

- If the victim is breathing, place him in the recovery position (see page86) and monitor for other problems. Assess the ABCs: Airway, breathing, and circulation.
- If necessary, start CPR (See section on CPR)
- Also, see section on first aid for hypothermia. Remove cold wet clothes from the victim and cover him with a blanket.

! WARNING: Continually monitor the victims ABCs and provide appropriate first aid until emergency services arrive.

Electrocution

Electrical shock may be a life-threatening emergency, depending on the degree of electrocution (how electricity was passed into/through the body and level of voltage to which the victim was exposed).

Electrocution Treatment

- First, ensure safe surroundings! Do not touch the victim. Electricity may still be flowing.
- Call for emergency services immediately.
- Turn off electrical source if possible. If not, move the source of electricity away from the victim using a non-conducting material such as plastic or wood.
- Assess the ABCs of first aid. If necessary, start CPR. Do not move the victim except to move him/her from dangerous surroundings.

See CPR, page 107, for additional information.

! WARNING: Continually monitor the victims ABCs and provide appropriate first aid until emergency services arrive.

Eye Injuries

There are several types of eye injuries. Many can be treated without a trip to the clinic or emergency room. Knowledge of appropriate first aid and monitoring for signs and symptoms of a more serious injury is essential.

First aiders should always ensure safe surroundings and use gloves and other personal protective equipment as necessary

Traumatic Blow to the Eye (Black Eye)

- Assess the injury, monitoring specifically for signs of a more severe injury:
- Any vision change, including blurry vision and double vision
- Severe pain
- Bleeding from the eye or blood in the colored or white part of the eye
- If any of these symptoms are present, seek professional medical care immediately

- Apply gentle pressure with a cold pack or ice wrapped in cloth. Do not apply pressure directly to the eyeball. Continue 15 minute cold pack treatments every 3-6 hours for 24-48 hours. Do not apply ice directly to skin.

Chemical Splash to the Eye

- The affected eye(s) should be flushed with water immediately. This can be done with lukewarm tap water from a shower or sink. Flushing should be done for 20 minutes. Victim and first aiders should wash hands and any other body parts potentially exposed to the chemical irritant.
- Do not rub the eye. This can cause further injury
- Remove contact lenses. Do not use contact lenses until symptoms have completely resolved
- Contact poison control center for additional advice. 1-800-222-2222
- If symptoms do not significantly improve following flushing, seek professional medical care immediately. Persistent pain, redness, swelling, and vision changes are particularly concerning symptoms.
- The injured person should wear sunglasses for a few days; the eye(s) will be sensitive to light and wind.

Foreign Particle in the Eye

- Examine the eye gently by pulling down on the skin below the eye to reveal the inner lower eyelid. Ask the victim to look upward as the lower eye is inspected. Upward pulling on the upper eyelid with the person looking downward can help inspection of the upper eye.
- If a foreign body seems to be floating on the surface of the eye or eyelid, flush the eye with saline solution or lukewarm water.
- Do not rub the eye
- Do not attempt to remove or dislodge embedded particles

If any of the following are present, seek medical care immediately:

- Particle embedded in the eye or particle cannot be removed by flushing the eye with saline solution or water.
- Victim has vision changes such as blurred or double vision
- Pain, redness, or swelling persists after particle has been flushed away.

Corneal Abrasion (Scratch on the surface of the eye)

- Perform inspection for foreign particles as described above.
- Saline solution or water can be used to rinse the eye.
- Do not rub the eye

- Do not touch the eyeball with fingers, cotton swabs, or other objects.
- Seek medical attention as soon as possible.
- Avoid use of contact lenses for several days
- Sunglasses may help with sensitivity to bright light and wind.

Cuts/Lacerations to Eye or Eyelid or Major Trauma to Eye

- Gentle pressure can be applied to the area around the eye. Do not apply pressure directly to the eyeball
- Call for emergency medical services or go to an emergency room immediately.

Fractures/Sprains/Dislocations

Injury to bones, ligaments, and tendons can range from mild bruising to complex fractures and soft tissue injuries requiring orthopedic surgery.

Most minor injuries can be treated with the R.I.C.E. method:

- Rest. The injured part should be rested until normal activity can be done without pain. Complete rest is not recommended. Gentle stretching and range of motion exercises can be started shortly after injury, as tolerated.
- Ice. Cold packs or ice water baths can be used, ideally starting shortly after injury and continued a few times a day for the first few days following injury. Ice treatments should not exceed 15-20 minutes and ice should not be applied directly to skin. Doing so can cause damage to skin and soft tissues.
- Compression. Use ad elastic bandage or neoprene brace. Compression dressings should be snug but not tight enough to restrict circulation.
- Elevation. Keep the injured limb above the level of the heart when possible.

Additionally, over-the-counter pain relievers such as acetaminophen, ibuprofen, or naproxen can be used (as directed on packaging) to help with pain.

More severe joint injuries may need additional care. Suspect a more serious injury in the setting of one or more of the following:

- An audible popping sound at the time of the injury
- Severe pain, inability to put any weight or force on the injured part.
- There is no improvement in symptoms after a few days.
- Symptoms of fractured or dislocated bone(s):
- The limb or joint appears deformed

- Severe pain with even slight touching of the injured area
- Bone piercing the skin
- Suspected severe ligament/tendon injury or fractures should be evaluated by a physician immediately.

911 or local emergency medical services should be called in the following situations:

- There is major trauma and multiple injuries are suspected
- The victim is confused or unconscious
- There is numbness or blue discoloration of body parts below the level of injury (fingers or toes, for example)
- Fracture is suspected in the head, neck, or back

Treatment for Suspected Fracture, Severe Ligament/Tendon Injury, or Dislocation:

- Assess the ABCs of first aid and treat accordingly
- Control any bleeding if there is an open injury (See section on Cuts, Scrapes, Bleeding)
- Immobilize the injured area. Minimize movement of the injured part. Do not attempt to reduce the fracture (realign the bones). Apply a splint if previously trained to do so.
- Apply a cold pack to the area to help with pain and reduce swelling. Do not apply ice directly to skin.
- If the victim is lightheaded and/or breathing rapidly (symptoms of shock), lay him down on his back and elevate the legs.

! WARNING: Continually monitor the victims ABCs and provide appropriate first aid until emergency services arrive.

Frostbite

Frostbite occurs when cold causes freezing of soft tissue, most commonly in the fingers, hands, toes, and feet.

Symptoms of frostbite include:

- Skin appears pale and feels hard, and waxy
- Area may be numb, person sometimes does not realize frostbite has occurred
- As tissue thaws, skin may become red and painful

Frostbite Treatment

- Get the victim out of the cold
- If possible, do not allow the victim to walk on frostbitten feet or toes.
- Place the affected part into warm (not hot) water. The water should be tested for comfortable temperature with unaffected tissue first.
- Alternatively, body heat can be used to thaw the tissue, for example, fingers can be placed in the underarm area.

- Do not rub or massage the area. This can cause further tissue damage.
- Do not use a fire, stove, heating pad, or heat lamp to thaw tissue. The numb tissue can be injured easily.
- Do not allow tissue to refreeze. If the victim will be exposed again to freezing temperatures, wait to thaw.
- Seek medical attention, even after tissue is thawed and symptoms improve.

Gunshot Wounds

Gunshot wounds are among the most severe and potentially fatal types of trauma. It is very difficult to assess the severity of a gunshot wound in the first aid setting. The size and location of bullet wounds are not always predictive of severity of injury. Appropriate first aid can help improve the chance of a nonfatal outcome, but death often occurs even with the best first aid.

Gunshot Wound Treatment

- As always, first ensure safe surroundings. Minimize the risk of additional injury by ascertaining whether the wound was accidental or intentional.
- First aiders should use disposable gloves and other personal protective equipment.
- Call for emergency services immediately.
- Do not move the victim unless absolutely necessary for safety or if he or she has to be transported to (or closer to) the emergency department by first aiders.
- Apply direct pressure to wounds to slow bleeding. Use towels, gauze, or if nothing else available, use hands to apply pressure. Do not remove pressure dressings from the wound. If dressings become soaked, add more gauze or cloth over the soaked dressings.
- Assess the ABCs of first aid (Airway, Breathing, Circulation). Begin rescue breathing or CPR if necessary. (See section on CPR for details).
- Place the victim in the recovery position (see page 86).
- Do not place a tourniquet on arm or leg with a gunshot wound unless bleeding cannot be stopped with direct pressure and risk of death from bleeding is high. (For additional information about tourniquets, see the section on amputations).
- Do not elevate the legs as a treatment for shock in a gunshot wound victim.

! WARNING: Continually monitor the victims ABCs and provide appropriate first aid until emergency services arrive.

Heart Attack/Chest Pain

A heart attack occurs when blood flow to heart muscle is stopped by blockage of a coronary artery. Irreversible damage to muscle tissue occurs quickly after its oxygen supply is cut off. As the saying goes, "time is muscle". Any time a heart attack is suspected, immediate medical evaluation is crucial. Important first aid steps can be taken before the victim arrives at the emergency department.

There are numerous causes of chest pain. Some causes (not just heart attack) are potentially fatal. We recommend all chest pain be evaluated by a physician immediately.

Symptoms of heart attack may include:

- Chest pain caused by a heart attack is variable. It may feel like pressure or squeezing inside the chest. It may be made worse with exertion and improve with rest.
- Radiation of chest pain. Pain may extend to the abdomen, arms, neck, jaw, or face.
- Shortness of breath
- Sweating, nausea, vomiting
- Dizziness, lightheadedness, fainting

Suspected Heart Attack:

- Immediate medical evaluation. Call 911 or local emergency services number. If phone access to emergency services is unavailable, transport the victim to the nearest hospital. Persons with chest pain should not drive themselves to the hospital.
- Aspirin. The victim should chew an aspirin while waiting for ambulance or on the way to the hospital, unless the person is allergic to aspirin or has been told by a doctor never to take aspirin.
- Nitroglycerin. Some persons with known heart disease carry nitroglycerin tablets. These can be used as directed by the person's physician. Never give nitroglycerin that is prescribed to someone else. This could make things worse.
- If the victim is unconscious, assess the ABCs of first aid. Start CPR if necessary.

See the section on CPR for additional information.

! WARNING: Continually monitor the victims ABCs and provide appropriate first aid until emergency services arrive.

Heat Related Illnesses

Heat related illnesses including heat exhaustion and heatstroke are occur when the body cannot keep itself cool. This is often related to heavy exercise in hot environments. Symptoms can range from mild muscle cramps to unconsciousness and coma.

Heat Cramps
Heat cramps are painful muscle spasms that can involve any muscles. The most commonly affected muscles, however, are calves, arms, back, and abdominal muscles.

Heat Cramps Treatment
- Ask the person to rest and cool off. Move the person to a shaded or air-conditioned area
- Give water or electrolyte-containing sports drink
- Do gentle muscle stretching and massage to affected areas
- Seek medical care if symptoms do not improve in 30-60 minutes

Heat Exhaustion
Symptoms often begin after heavy exercise in hot environments. Signs and symptoms may include:
- Fatigue, lightheadedness, and/or dizziness
- Headache
- Heat cramps
- Rapid heart rate, low blood pressure
- Elevated temperature

Heat Exhaustion Treatment
- Ask the person to rest and cool off. Move the person to a shaded or air-conditioned area.
- Ask the person to lay down with feet elevated.
- Give water or electrolyte-containing sports drink.
- If a thermometer is available, check temperature
- If the person loses consciousness (faints), is becoming more confused, or has a temperature above 102 F (38.9° C), call emergency medical services immediately.

Heatstroke
Heatstroke is the most severe of the heat-related illnesses. Like heat cramps and heat exhaustion, it is usually associated with heavy exercise in hot surroundings, often in the setting of inadequate fluid hydration. Young children, elderly persons, and chronically ill persons are at higher risk of heatstroke. Alcohol use is also an important risk factor for heat-related illnesses. Signs and symptoms may include:
- Elevated temperature.
- Rapid heart rate and breathing rate
- Nausea and vomiting
- Abnormally high or low blood pressure
- Skin may be moist or dry (if body is unable to produce sweat)

- Dizziness, lightheadedness, or fainting
- Feeling worried or anxious
- Headache, irritability, confusion

Suspected Heatstroke

- Move the person to a shaded or air-conditioned area.
- You may spray mist cool water on the person or cover him with a damp towel or sheet. You may fan cool air toward the patient.
- Call for emergency medical services.
- If the person can safely swallow, give water or sports drink.

! WARNING: Continually monitor the victims ABCs and provide appropriate first aid until emergency services arrive.

Hypothermia

Hypothermia is defined as core body temperature less than 95 degrees Fahrenheit (35 degrees Centigrade). Hypothermia can cause a wide range of symptoms from mild chills to loss of consciousness, coma, and death.

Hypothermia often occurs when victims are inadequately prepared for cold weather conditions. Common contributing factors include insufficient clothing and wet clothing. Cold, windy, and wet conditions can lead to rapid loss of body heat.

Symptoms of hypothermia include:

- Cold, pale or red skin
- Shivering
- Fatigue
- Confusion
- Slurred speech
- Shortness of breath
- Slow breathing

Suspected Hypothermia:

- As always, first ensure safe surroundings. First aiders are also at risk of hypothermia. Move victims out of cold conditions. If indoor shelter is not immediately available, cover the victim's body and head and protect him from the wind.
- Call for emergency medical services.
- Evaluate the ABCs (airway, breathing, circulation). If breathing stops or becomes severely slow, start CPR (See section on CPR)
- Remove wet clothing and replace with dry covering.
- Do not apply direct heat (such as hot water, heat lamp, or fire) to the body. You can use warm compresses to the chest, neck, and groin.
- Do not attempt to rewarm arms and legs first! This can

encourage circulation of cold blood back to the body (brain, heart, lungs), which can cause further drop in core body temperature and worsening symptoms!
- Do not give the victim alcohol. If the victim can safely swallow, give warm (not hot) fluids such as tea, coffee, or hot chocolate.
- Stay with the victim and monitor for worsening symptoms until emergency medical services arrives.

! WARNING: Continually monitor the victims ABCs and provide appropriate first aid until emergency services arrive.

Influenza /Flu

Influenza is a viral infection of the respiratory tract. It is highly contagious. Symptoms can vary from mild common cold symptoms to severe respiratory compromise including respiratory failure and death. There are numerous types of influenza including the 2009 H1N1 strain that caused numerous recent outbreaks worldwide.
First aid for influenza consists mainly of assessment of the severity of symptoms, supportive care for the infected person, and prevention of spread of infection.

Symptoms of influenza infection

- Fever, often above 101 F (38.3 C)
- Lethargy
- Cough
- Sore throat
- Runny or stuffy nose
- Headache
- Chills
- Body aches
- Nausea and vomiting

Supportive Care for Suspected Influenza:

- Encourage the infected person to drink plenty of fluids
- Treat fever, aches, and cough with over-the-counter medications such as acetaminophen, ibuprofen, and cough syrup. Follow the dosing instructions on the packaging.
- Prescription drugs to treat influenza are helpful if started within the first 48 hours of the start of symptoms.
- Encourage hand washing and covering coughs and sneezes. The infected person should rest at home until feeling better and fever has resolved for at least a full day.

Seek Medical Care Immediately For Any of the Following:

- The sick person is an infant or child or is elderly or has chronic medical problems
- Difficulty breathing
- Vomiting with inability to drink plenty of fluids
- Dizziness, confusion

Lightning Strike

- Seek medical help immediately.
- Assess the situation. Is it safe to help the victim(s)? Do not risk more casualties. If necessary, move the victim to a safer location before providing first aid.
- Check ABC's
- If not breathing and no heartbeat, immediately begin CPR, and continue until rescue arrives. (If not responsive after 30 minutes, the chances of survival are slim.)

Lightning Related Injuries

In addition to cardiac and respiratory arrest, other lightning-caused injuries are burns, shock, brain injury, muscular and skeletal damage, blunt trauma injuries including broken bones and ruptured organs. Treat all symptoms until help arrives.

! WARNING: Continually monitor the victims ABCs and provide appropriate first aid until emergency services arrive.

Nosebleeds

Nosebleeds occur commonly. Most often, the source of the bleeding is just inside the nose; however, sometimes the blood comes from deeper inside and can be difficult to stop, occasionally requiring physician evaluation.

Nosebleed Treatment

- Use your fingers to pinch the nose just above the nostrils
- Sit down, lean forward and breath through your mouth
- Remain in this position for 5 to 10 minutes.
- If bleeding continues, you may use a decongestant nasal spray (such as Afrin or Neosynephrine) in each nostril, followed by repeating 10 minutes of nostril pinching. If bleeding continues for more than 20 minutes, seek medical attention
- Medical attention should be considered sooner with nosebleeds caused by trauma and in persons taking blood-thinning medicines.

Prevention of nosebleeds caused by drying and cracking can be improved by keeping the nose moist with nasal saline spray or water-based nasal lubricants.

Poisoning / Intoxication

Serious illness or death can result from accidental or intentional intoxication from thousands of different drugs, chemicals, and medications. As may be suspected, the large number of potential poisons can cause a very wide range of symptoms (in addition, many toxins cause no obvious symptoms at all).

Common possible symptoms will be reviewed; however, whenever poisoning is known or suspected, a poison control center and/or emergency medical services should be called immediately.

Signs and symptoms of poisoning:

- Victim with altered mental status (confusion, slurred speech, unconsciousness)
- Chemical odor on breath

Call for Emergency Medical Services Immediately If:

- The victim is unconscious or has altered level of consciousness
- Victim is not breathing or breathing very rapidly or very slowly.
- Severely agitated or having seizures.

If the victim appears to have a normal level of consciousness and is breathing comfortably, but poisoning is suspected:

- Call a poison control center immediately. In the United States, the phone number is 1-800-222-1222.
- Be prepared to give information about the victim's symptoms and the chemical in question (if known)
- Help the victim remain calm. Keep him/her in a comfortable position in a well-ventilated area.
- Follow the instructions given by the poison control center.
- Do not induce vomiting with ipecac syrup or any other method. This practice is no longer recommended.
- If advised to go to the hospital, bring the medication or chemical containers for physicians to examine.
- Reassess the victim frequently for changes or new symptoms.

! WARNING: Continually monitor the victims ABCs and provide appropriate first aid until emergency services arrive.

Puncture Wounds

Puncture wounds often leave a relatively small skin defect. The deeper tissues, however, are the sites of potential injury or infection.

For puncture wounds caused by bites, thorns, or splinters, (See the first aid instructions in the section on animal bites). First aid comprises managing bleeding, cleaning and dressing the wound and monitoring for signs of infection. Tetanus booster may be necessary.
Information on first aid for other types of puncture wounds can be seen in the sections on: penetrating chest wounds, abdominal injury, and cuts/scrapes/bleeding.

Seizures

Seizures are the result of sudden abnormal electrical and chemical activity in the brain. there are many types of seizures and symptoms can range from abnormal movements or sensation of one body part (partial seizure) to full loss of consciousness and muscle shaking/convulsions (grand mal seizure). There are numerous possible causes of seizures, including:

- Epilepsy (a disorder that may cause recurrent unprovoked seizures)
- Fever (these seizures are known as febrile seizures)
- Several medical conditions including stroke, head injury, medication toxicity, eclampsia (during pregnancy), alcohol withdrawal, and many others.

Seizures First Aid

In most cases of seizure, the mainstay of emergency aid revolves around keeping the person safe from self-injury.

- Move the person away from sharp edges
- Place the person in the recovery position (see page 86).
- Place something soft (a pillow or blanket) beneath the head.
- Do not try to restrict movements or stop convulsions by holding the person down.
- Don't place anything into the person's mouth. Contrary to common belief, it is not possible for people to swallow their own tongue.
- If the person begins vomiting, allow the material to flow out of the side of the mouth on its own.
- Call emergency services

After a seizure, it is common for a person to be very sleepy, confused, and exhausted. This is known as the post-ictal state. It can last minutes or even hours. As a person recovers, she may not know she just had a seizure. It is important to support and reassure her until normal state of consciousness returns.
An infant or child with a febrile seizure may be sponged with lukewarm water, but should not be immersed in water.
! WARNING: Continually monitor the victims ABCs and provide

appropriate first aid until emergency services arrive.

Stabbing

See Cuts (page 115), Penetrating Chest Wounds (page 112), Abdominal (page 87) and Eye Injuries (page 119), depending upon location.

Stroke CVA/Tia

A stroke occurs when part of the brain's blood supply is blocked by a clogged artery or if there is bleeding into the brain. Once deprived of oxygen, brain cells begin to die in a matter of minutes.

A stroke is a true medical emergency! If stroke is suspected (See below), call for emergency medical care immediately. Every minute is crucial to minimize permanent brain damage. Unless CPR is necessary, there's not much one can do for someone having a stroke—no aspirin! Get patient to the hospital!

The National Institute of Neurological Disorders and Stroke (NINDS) reports the following as common symptoms of stroke. If any of the following occur, call emergency services or go to the emergency room immediately:

- Sudden numbness or weakness in the face or limbs, often on one side of the body
- Difficulty speaking or loss of ability to speak.
- Sudden changes in vision (blindness, blurring, dimming) in one or both eyes
- Sudden severe unexplained headache, confusion, dizziness, or falls may accompany the symptoms listed above.

! WARNING: Continually monitor the victims ABCs and provide appropriate first aid until emergency services arrive.

Vomiting

See Section on Diarrhea, Nausea, and Vomiting, page 117.

Notes:

Lightning: Precautions for Workers

If you work in high risk locations, you can take steps to protect yourself.

Check the Forecast
Know the daily weather forecast so you are prepared and know what weather to expect during the day.

Watch for Signs
Pay attention to early weather signs of potential lightning strikes such as high winds, dark clouds, or distant thunder or lightning. When these occur, do not start any activity that you cannot quickly stop.

The 30-30 Lightning Safety Rule
After you see lightning, if you cannot count to 30 before hearing thunder, go indoors. Suspend activities for at least 30 minutes after the last clap of thunder

Follow the Company Program
Know your company's lightning safety warning program, if it has one. These programs should include access to a safe location and danger warnings that can be issued in time for everyone to get to the safe location.

Assess the Threat
Although no place outside is safe during a storm, you can minimize your risk by assessing the lightning threat early and taking appropriate actions. For example, if you hear thunder, lightning is close enough to strike you. Stop what you are doing and seek safety in a building or metal-topped vehicle with the windows up. Also, try to follow the 30/30 rule for lightning safety.

Avoid tall Structures
Avoid anything tall or high, including rooftops, scaffolding, utility poles, ladders, trees, and large equipment such as bulldozers, cranes, and tractors.

Avoid Conductive Materials
Do NOT touch materials or surfaces that conduct electricity, including metal scaffolding, metal equipment, utility lines, water, water pipes, or plumbing.

Stay away from Explosives
If you are in an area with explosives, leave immediately.

IMPORTANT REMINDER: If your coworker is struck by

lightning, he or she DOES **NOT** carry an electrical charge. Call 911 and immediately begin first aid response if necessary.

Lightning: First Aid Recommendations

How to Help

Giving first aid to lightning strike victims while waiting for professional medical attention can save their lives. It is safe to touch a lightning strike victim. People struck by lightning DO NOT carry a charge.

Follow these four steps immediately to help save the life of a lightning strike victim:

Call For Help

Call 911 immediately. Give directions to your location and information about the strike victim(s). It is safe to use a cell phone during a storm.

How many victims are there?
Where was the victim struck?
Is the storm still continuing?

Assess The Situation

Safety is a priority. Be aware of the continuing lightning danger to both the victim and rescuer. If the area where the victim is located is high risk (e.g., an isolated tree or open field), the victim and rescuer could both be in danger. If necessary, move the victim to a safer location. It is unusual for a victim who survives a strike to have any major broken bones that would cause paralysis or major bleeding complications unless the person suffered a fall or was thrown a long distance. Therefore, it may be safe to move the victim to minimize possible further exposure to lightning.

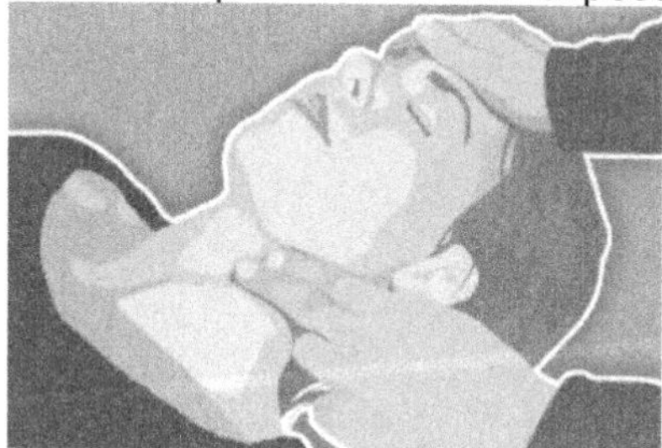

Respond

Lightning often causes a heart attack. Check to see if the victim is breathing and has a heartbeat. The best place to check for a pulse is the carotid artery which is found on your neck directly below your jaw, as shown in the picture.

Resuscitate

If the victim is not breathing, immediately begin mouth-to-mouth resuscitation. If the victim does not have a pulse, start cardiac compressions as well (CPR). Continue resuscitation efforts until help arrives. If the area is cold and wet, putting a protective layer between the victim and the ground may help decrease hypothermia (abnormally low body temperature).

IMPORTANT REMINDER: Lightning may also cause other injuries such as burns, shock, and sometimes blunt trauma. Treat each of these injuries with basic first aid until help arrives. Do not move victims who are bleeding or appear to have broken bones.

COMPANY LIGHTNING PROGRAM NOTES:

HELICOPTER DITCHING

Helicopter/Aircraft Ditching

Be Prepared

Most Ditchings occur in critical phases of flight - Take Off, Landing or Hover.

- 92% have less than 1 minute warning.
- 28% have less than 15 seconds warning.

Mitigate

- Be Equipped w/ Life Jackets (PFDs) & Raft / Exposure Suits
- Wear PFDs over Water
- Have a Plan
- Practice your Plan

Brace Positions

Double Strap Restraint Systems

- Keep feet outside of seat crush zone, forward of seat and flat on floor.
- Cross arms.
- Slip thumbs under shoulder harness straps.
- Grip straps firmly.
- Tuck head into the V formed by your crossed arms.
- This will help prevent your neck from rotating forward and hyper extending.
- Seat belts should be low on the hips and as tight as possible.
- Shoulder restraints should be tightened as much as possible.
- Seat should be aft as far as possible.

For single strap shoulder restraint systems

- Keep feet outside of seat crush zone, forward of seat and flat on floor.
- Cross arms.
- Slip thumb under shoulder harness strap.
- Grip strap firmly.
- Grasp your shoulder with the other hand.
- Again, this forms a V in which you nest your head.
- Then tuck your head into the V formed by your arms,
- Grip the shoulder strap and your unrestrained shoulder very tightly.

Egress -Time to Get Out

- Establish and Hold Reference Point
- Keep your feet on the deck to maintain orientation.
- Remember - what was on your right when you were upright is still on your right when you are inverted.
- Do not release restraints 'till motion stops!

- Don't let go with both hands at the same time!
- Open Doors - Windows
- Wait for Motion to Stop
- Take a Deep Breath before being submerged.
- Count 3 - 4 seconds - release harness
- Use Hand over Hand method to Egress- always have one hand in contact w/ the aircraft to remain oriented.
- **DO NOT INFLATE PFDs until clear of aircraft!**
- Breath out - bubbles go to surface
- Get Clear of Aircraft
- Secure raft to yourself, not to airplane. Tie individual rafts together
- You may have less than a minute before aircraft is submerged
- A Seat Belt Cutter may be a useful tool to have readily available. They are inexpensive, and could save your life if your restraints do not release.

Survival

- Get Away from Aircraft
- All PFD's Inflated
- Do a Head Count
- Deploy Raft - Get In
- Inventory Gear - Assess Situation
- Remain afloat – Life Jacket / PFD
- Get out of the Water - Raft or Immersion Suit
- Get help – Signaling Gear, PLB

Cold Water is a Big Hazard

- You have to get out of the water, or stop the heat loss, or **you <u>will</u> die.**
- The clock is running . . .
- Your remaining lifespan depends on the temperature of the water and how you can stop your heat loss.
- Hypothermia can begin within 10-15 minutes.
- Hypothermia can cause death, or contribute to drowning.
- Unconsciousness occurs when core temp. is 89.6 degrees. (Normal 98.6)
- Death likely when core cools below 86 degrees.

Under good conditions

(life jacket, light clothing, staying still) --

- 60 degree water - survival time 7 hours
- 50 degree water - survival time 2.5 hours
- 40 degree water - survival time 2 hours
- 32 degree water - survival time 1.5 hours

Survival Factors in Cold Water

- Will to Live - Most important in all survival situations.
- Flotation - Personal Flotation Device (PFD) essential.
- Heat Retention - Clothing / Raft / Survival Gear

"STAY" Rules for Cold Water Survival

- Stay Afloat

- Stay Dry
- Stay Still
- Stay Warm
- Stay with Aircraft / Boat

Stay Afloat

- Must breathe to prevent drowning
- Must control panic to breathe.
- Panic decreases ability to float.

Lifejacket / PFD

- Non-swimmers need assistance of PFD.
- Provides advantage recovering from cold shock and allows better breath control.

Without PFD

- Flotation is possible even with heavy clothes.
- Trapped air in clothing assists flotation.
- Hold onto floating debris.

Stay Dry

- Get out of water ASAP.
- If that's impossible, get main heat loss areas out of water (hang on to floating object).
- Get head dry and out of water.
- Head in water increases heat loss by 80% over head out of water.
- A dry suit is best protection, but not as good as being out of the water.

Stay Still

- Movement increases circulation and heat exchange in extremities.
- Staying still decreases heat loss by 30% over swimming or treading.
- It is difficult to float motionless with out Lifejacket / PFD

Stay Warm

- Main Heat Loss Areas
  - Head & Neck
  - Groin
  - Sides of Chest
- Protect main heat loss areas
- Wear coat & hat
- If getting out of water is impossible, assume HELP, HUDDLE, Human Carpet or Human Chain positions.
- These positions double survival time over swimming or treading.
- These positions are **impossible** without a PFD.

H.E.L.P.
Heat Escape Lessening Posture
Impossible without a PDF

HUDDLE
A "group hug" to conserve heat
Impossible without a PFD

Stay with Aircraft / Boat

- May be possible to get out of water.
- Better chance of being spotted - larger target.
- Success in swimming to shore depends on many variables. Swimming increases heat loss.
- In 50 degree water, average person wearing PFD and light clothing can cover a distance of only .85 mile before being incapacitated by hypothermia.

Signaling Devices

- Mirror
- Flares
- Whistle
- ELT (or PLB can be carried as extra equipment)
- Dye – See Rescue Device
- Chemical Light Sticks
- Strobe
- Cell Phone or Aviation Handheld Radio if in waterproof bag

Practice your Plan

- Make Ditching / Egress procedures part of every pre-flight briefing.
- Include:
- Emergency calls
- Ditching procedures
- Brace Positions
- Removal of restraints
- Egress procedures
- Survival equipment

Brooks' Golden Rules how to survive a helicopter crash/ditching in water.

The latest paper published in 2014 on survival from a helicopter accident in water (133 helicopters) reveals that approximately 25% of crew and passengers do not survive (reference). These statistics have remained unchanged for the last 30 years. So if you wish to survive you better pay attention to some simple rules.

Accidents occur during all phases of flight, take off, cruise, and landing. In the majority of cases you will have less than 15 seconds of warning before you find yourself in the water. This is not good news. So you must be mentally and physically prepared for a ditching throughout the entire flight.

Before you step on to the helicopter, you will be issued with an EBS. Make sure that the mouthpiece is clean and ready to be deployed, make sure the bottle is full and the valve to turn it on is switched to the 'ON' position. There will be a 'Press to test' button on the EBS. Some crew and passengers like to give it a press. This gives them confidence that the system is working. But, if you do this, only give it a very short press; remember this is the air that you need if you have to make a complex underwater escape. Don't waste it.

As you step on board and make your way to your seat, have a look around for obvious obstructions and places where you can get snagged e.g. internal fuel tanks.

Always try to sit next to an exit in a window seat.

Once sat down before you do up your harness, take a look at the window/door, the jettison method (pull tab, lever, push out etc.). Create a mental and physical schema in your brain by feeling for the window sill, the mechanism and or the place where you have to push to jettison the window; note how far you have to reach. Also, recycle your harness release mechanism.

When you are strapped in, make sure that your suit is zipped up tight, that your life jacket and EBS are fitting correctly and your personal locator beacon is secure. Again create a mental and physical schema by feeling for the toggles on your life jacket, the mouthpiece on your EBS, and the location of your gloves. Make sure that your mike cord is not jammed behind the head rest when you put your seat back backwards and forwards, and take a look around the cabin to see which secondary exit you may have to use if your primary exit is jammed.

Stow all your gear safely. When the water comes in, one pilot described it to me like being hit in the body with a fire hose. So anything unsecured will be violently washed around the cabin, and YOU are the human target on the receiving end.

If you are sitting in an aisle seat, again make a mental and physical schema of which route you are going to take to escape. If the person in the aisle seat has been seriously injured, he/she will not be able to jettison the exit and may even jam it. So you will have to go cross cabin. You have not time to wait, you must go.
Assume the best crash position that you can for whatever type of harness that is fitted (not easy in the bulky survival suits), try to make your body profile as small as possible so that you are less of a target for debris flying around the cabin and this reduces the likelihood of you getting disorientated.
Be prepared for cold shock, even though you theoretically are wearing a leak tight suit, very cold water on the face may precipitate the gasp reflex and hyperventilation, so have your EBS ready.
Should you jettison the exit just before landing on the water, just after you have landed on the water, or wait for the helicopter to come to a stable floating position (this may be inverted, and submerged)? Because the majority of accidents occur with little warning, this may only be an academic question. I cannot give advice on this because we just do not have enough evidence which is the best procedure; you will just have to make your own decisions on this.
The more practice you can have in the dunker the better your chances, so rather than wait until your refresher course is due and wait to the bitter end to sign up for a new course, be proactive and do it a few months earlier. At present (2014), EBS training is only taking place in the SWET chair, volunteer to use it in the dunker.

If you really want to be pro-active, then if you take a daily cold shower for three consecutive weeks, that will protect you from cold shock for around six months.
EXPECT TO ESCAPE UNDERWATER AND UPSIDE DOWN.
Good luck.
Reference.
Brooks C.J. Macdonald C.V. Baker S. Shanahan D.F. Haaland W.L. Helicopter crashes into water: Warning time, Final position, and other factors affecting Survival. Aviation, Space and Environmental Medicine (2014);85:440-4.
Dr. Chris Brooks, Kanata, Ontario, K2T1K2. April 2114.

Thanks to Dr. Chris Brooks

SEA STATE SCALE

SS0

Sea like a mirror; wind less than one knot. Average wave height is 0.

SS1

A smooth sea; ripples; no foam; very light winds, 1-3 knots, not felt on face. Average wave height is from 0-0.3 m (0-1 ft).

SS2

A slight sea; small wavelets; winds light to gentle, 4-6 knots, felt on face; light flags wave. Average wave height is 0.3-0.6 m (1-2 ft).

SS3

A moderate sea; large wavelets, crests begin to break; winds gentle to moderate, 7-10 knots; light flags fully extend. Average wave height is 0.6-1.2 m (2-4 ft).

SS4

A rough sea; moderate waves, many crests break, whitecaps, some wind-blown spray; winds moderate to strong breeze, 11-27 knots; wind whistles in the rigging. Average wave height is 1.2-2.4 m (4-8 ft).

SS5

A very rough sea; waves heap up, forming foam streaks and spindrift; winds moderate to fresh gale, 28-40 knots; wind affects walking. Average wave height is 2.4-4.0 m (8-13 ft).

SS6

A high sea; sea begins to roll, forming very definite foam streaks and considerable spray; winds a strong gale, 41-47 knots; loose gear and light canvas may be blown about or ripped. Average wave height is 4.0-6.1 m (13-20 ft).

SS7

A very high sea; very high, steep waves with wind driven overhanging crests; sea surface whitens due to dense coverage with foam; visibility reduced due to wind blown spray; winds at whole gale force, 48-55 knots. Average wave height is 6.1-9.1 m (20-30 ft).

SS8

Mountainous seas; very high-rolling breaking waves; sea surface foam covered; very poor visibility; winds at storm level, 56-63 knots. Average wave height 9.1-13.7 m (30-45 ft).

SS9

Air filled with foam; sea surface white with spray; winds 64 knots and above. Average wave height is 13.7 m and above (45 ft and above).

COMMERCIAL AIRCRAFT SAFETY

! Disclaimer: this information is advisory in nature and is not intended to identify all scenarios or situations a person might encounter.
! Following these guidelines will not guarantee your safety.

Note: This information should be read and understood immediately. During an emergency is no time to try and figure out what to do. Regular review and practice will help you prepare for an emergency.

What to Do In Case of an Aircraft Emergency

Although the U.S. airline system is the safest in the world, crashes do occur. However, nearly all crashes have survivors. The tips below, courtesy of the FAA, can help ensure that you survive a crash (which causes 10 percent of airline deaths) and the resultant fire and smoke (which causes the other 90 percent). Do not depend upon others. You are your own safety officer. Survival favors the prepared.

Airport Dress

- Dress casually (slacks, no tight fitting clothes, no skirts) in case you have to climb over obstacles to leave the plane.
- Wear natural fibers. Synthetic clothing, including nylons, burns right through the skin causing severe injury. If you are wearing nylons and have to slide down the emergency chute, the friction could melt the material into your skin.
- Wear bright colors. You can be seen better if you need emergency treatment outside the plane. If you collapse on the ground, you will hopefully not be run over by an emergency vehicle.
- Do not wear high-heeled shoes; they could puncture the exit chute.
- Do not wear pierced earrings. The safety vest inflates above the ears. The earrings could puncture the safety vest, losing 50 percent of buoyancy. If you are in cold water, the vest keeps your head above water, helping to retain your body heat. Losing the buoyancy of the vest dramatically increases the chance of body heat loss and death from hypothermia.
- Wear laced shoes and keep them on during takeoff and landing. If preparing for a crash, put your shoes back on. In a crash, loafers may fly off from the G-forces. Avoid walking where there might be debris such as glass, razor-sharp metal shards, or fuel. Also avoid touching the cabin if there is a fire as the metal would be hot.
- Because the plane is set to a low humidity (between 4 and 15 percent) you dehydrate while in the air. Drink plenty of water or juice at home and before boarding the plane. While in-

flight, drink fluids, even if you are not thirsty. Dehydration parches your throat and nasal passages, which will have a hard enough time from the smoke soot. A word of caution: alcohol speeds dehydration.
- Do not take any medication that may slow think thinking and reaction time in an emergency (i.e., sedatives) unless prescribed by a physician. Regarding prescription medication, if you are traveling in different time zones, make sure you take your medication according to the number of hours between doses, not by the time on your watch. There is a good chance you could either overdose or underdose.

Once In the Aircraft

- Where to sit: The best place to sit is either on an exit row or within two rows of one. Most people instinctively exit a plane the way they entered. Make sure you know where the closest emergency exits are. Those sitting in exit rows are crucial to everyone's safety. Make sure that those sitting on the exit rows speak and understand English. The FAA requires that they be able-bodied enough to remove the window (it weighs 40-70 pounds) or open the door. If you notice these rules not being followed, you have the right (and obligation) to report the situation to the flight attendants to arrange for a passenger to move to another seat.
- Removal of the emergency exit window is an important first step in crash survival.
- Be grateful for the tight leg room. It is safer because there is less room to be thrown around.
- Check with the air carrier regarding the number and size of carry-on bags. Put a softer, lighter bag (with no sharp edges) in the top bin. In an emergency, these bins pop open (they are rated for only 3 Gs) and contents become projectiles. A heavier bag should be placed under the seat in front of you. In case of an emergency while the plane is still moving, brace your feet against the bag to keep it from traveling under your feet where you might trip in your haste to leave the plane.
- While on the airplane, remember to keep items like your laptop computer near your seat and not in an overhead compartment away from your view.
- Pay heed to the flight attendant's emergency instructions. All planes (even the same models from the same manufacturers) are configured differently, particularly regarding the location and operation of emergency doors and window. Know where the nearest two exits are; doors can jam because of a crash. Count the number of rows you are away from these exits. When the plane fills up with smoke, visibility is zero. Back up what the attendant says by reading the emergency card in

the flap in front of you. Caution: Look before you reach into the pocket.

- Passengers have been stabbed with discarded hypodermic needles.
- Eighty percent of all accidents happen at takeoff and landing. Make sure you are buckled up securely as acceleration and deceleration causes the body to lurch forward and backward, which could cause injury. Never release your seatbelt until the plane comes to a complete stop.
- Keep the seatbelt buckled when seated. Most injuries from air turbulence occur in a split second. One hundred percent of the injuries could be eliminated if seatbelts were worn.
- Keep debris off the floor, especially magazines with slick covers, which could cause you to slip when in a hurry.

Should the Worst Happen

- Once the plane comes to a halt, what you do in the first 90 seconds may decide your fate; knowledge of your surroundings is crucial. Never release your seatbelt until the plane comes to a complete stop and you have observed your surroundings. If you find yourself upside down, releasing your seatbelt could prove hazardous.
- The seat cushion floats do not work very well in water because they are unstable and force you to use energy to stay up. Use them until you find a better alternative, such as the exit chute that can serve as a raft. The inflatable safety vests are also good bets as they keep your head above water even if you are unconscious. When making reservations, you should ask the airline for flotation devices for children and infants.
- If traveling with your family, get seats next to each other. Do not leave the lives of loved ones in the hands of panicky strangers.
- Before removing an exit door or window, make sure you see no fire outside. You court disaster by allowing the fire and smoke to draft inside.
- If the window must be removed, sit down to do it. If you stand, the person behind you will be pushing you and the window cannot be brought inside before it releases. Your knees can block panicky passengers until you can move the window. The airlines' placards instruct you to place the removed exit window on the seat (it saves money). The best thing to do is to pull to release, rotate, then throw the window out the opening to get it out of the way.
- Do not come out head first, unless it is a water landing. You could be pushed out, landing on your head. First put out a leg, then your body, then your other leg, thereby maintaining your balance.

- If the plane breaks apart, consider using the new holes as exits.
- There is an emergency rope in the cockpit. If it is the only way out, close the door behind you to block out the smoke, pop out the window, and climb down. This is not the best way to leave the plane, but it might be your only way.
- If the exit chute does not deploy, reach down and pull the handle at the base of the door jamb.
- The steeper the chute, the faster you travel.
- Jump feet first into the center of the slide; do not sit down to slide. Cross your arms across your chest, elbows in, with legs and feet together or crossed. If you try to brace yourself with your hands while traveling downward, severe friction burns can occur.
- There are no exit chutes over the wings on some domestic flights. The pilot will bring the flaps down to enable people to slide off the wing to safety.
- Leave belongings behind. Do not risk your life and the lives of others by slowing down to retrieve things. Do not carry bags out–if they get stuck, even for a few seconds, you are dooming those behind you.
- Move away from the aircraft, fire, and smoke. If possible, help those requiring assistance. Never go back into a burning aircraft.

Remain alert for emergency vehicles.

Notes:

GRADE CROSSING SAFETY

GRADE CROSSING SAFETY PROCEDURES

- When you see an Advance Warning sign, it alerts you to a railroad crossing ahead. It is time to begin to slow your vehicle, so you will be able to stop if a train is approaching.
- While slowing or stopped, look and listen carefully in each direction for the sight and sound of a train.
- Never shift on a railroad crossing to avoid the risk of stalling on the tracks.
- Make sure that trailer jacks are in the up position; non-retracted trailer jacks can cause trailers to become stuck on crossings.
- Check for traffic around you before you start to move towards a crossing. Use a pull-out lane, if one is available. Turn on your flashers, if necessary to warn traffic that you are slowing down or stopping at the crossing. At crossings, don't stop any closer than 15 feet. If you're in traffic, don't go forward if you can't safely clear the crossing.
- Don't start across until you know you can cross the tracks completely without stopping.

BEFORE RESUMING TRAVEL

- Take a quick look in both directions before you start your rig across.
- If there is a traffic signal or a stop sign across the tracks, make certain traffic will not trap you on the crossing.
- Before you cross, plan to have 15 feet clearance between your ICC (rear) bumper of your truck and the farthest rail. This will prevent your truck's overhang from getting hit. Keep in mind there's an overhang, both for your truck and a train, of 3 feet or more.
- If there are flashing lights and gates at the crossing, stop when the lights start to flash. Wait until the lights stop flashing and the gates go completely up.
- If there is no gate, but warning lights are flashing, you will be required to stop, then can proceed when it is safe to do so.
- If the warning lights at the crossing begin to flash *after* you have started across the tracks with your rig, keep going. Do not back up.
- If you get stuck at the crossing, get out and call the 800 number posted at the crossing, or call local police to alert trains of your position.

Railroads' Emergency Phone Numbers

Use these phone numbers to report a vehicle stalled or hung up on tracks, or a signal malfunction. Provide the location, crossing number (if posted), and the name of the road or highway y that crosses the tracks. And be sure to specify that a vehicle is on the tracks!

Amtrak 1-800-331-0008
BNSF Railway 1-800-832-5452
CSX 1-800-232-0144
Canadian National 1-800-465-9239
Canadian Pacific 1-800-716-9132
Kansas City Southern 1-877-527-9464 or 1-800-892-6295
Norfolk Southern 1-800-453-2530
Union Pacific 1-888-877-7267

Call the local police or 911 if you cannot locate the railroad emergency phone number at the site.

VEHICLE ACCIDENT W/ RADIOACTIVE MATERIAL

#1 Help Injured Individuals

The likelihood that nuclear radiation levels at an accident scene would be high enough to cause injury is extremely remote. Unbroken radioactive material packages never have a surface radiation level exceeding 1000 mR/hr, and packages likely to break under accident conditions are used only for materials low enough in radioactivity to present no immediate hazard if dispersed. Therefore, help for injured individuals should not be delayed out of concern for radiological hazards. The person should give normal first aid to the extent qualified. If a radioactive materials package has been badly damaged or if you suspect that it is leaking, do not panic.

The steps to take are simple:

1. Stay away from the package and do not touch it.
2. Keep other people away from the package.
3. Tell anyone who may have touched the package to remain on-hand to be checked by radiation protection specialists.
4. If you touched the package or objects near it, wash your hands with lukewarm water.

#2 Notify the Authorities

Using any form of communication available, an individual involved in an accident should notify the authorities of the accident. The local 9-1-1 emergency service or the local police or fire departments should be able to respond properly.
It is important to give the greatest amount of detail possible when calling for help. Important information includes:

The location and nature of the accident.
The cargo (if easily identified by vehicle placards or package labels).
Your name and the phone number from where you are calling (if applicable).
The number of persons injured and the seriousness of their injuries.
The actions being taken at the time of the call.

*If the emergency operator requests, communications should be maintained until the authorities have arrived. It is important for those first on the scene to wait for the arrival of authorities and give them a full description of the events that occurred before their arrival. This will ensure that all persons involved understand the potential hazards and that all personnel will receive proper medical treatment and be decontaminated as required.

#3 Isolate the Area

Once injured individuals have been helped and the authorities have been notified, the accident scene should be isolated. Two reasons are:
To prevent the spread of low-level radioactive contamination.
To prevent exposure to high-levels of radiation in the highly unlikely event of a release of highly radioactive materials or a high level sealed source.

*Radioactive materials released at an accident scene, even at levels of little consequence, can result in very small but still detectable levels of contamination being spread a great distance. The spread of contamination can be controlled by limiting access to and egress from the accident scene. Although, in some cases, the contamination spread would be of insignificant radiological consequence, any detectable amount can prove to be of great concern to the public and news media.
It is important to treat everything that has been near the accident as potentially radioactive and contaminated until it has been verified by qualified radiation protection personnel to be free of radioactive contamination. Individuals who have contacted potentially contaminated materials should remain on-hand until they have been checked by qualified personnel. **Only qualified personnel should attempt to clean up a spill of any hazardous materials--radioactive or not.**

Emergency Contacts – Company Specific Notes

SAFE CRANE WORKING DISTANCES FROM POWER LINES

Safe working distance from power lines.

a. When operating near high-voltage power lines:	
Normal voltage (phase to phase)	Minimum required clearance
to 50 kV	10 ft (3.1 m)
Over 50 to 200 kV	15 ft (4.6 m)
Over 200 to 350 kV	20 ft (6.1 m)
Over 350 to 500 kV	25 ft (7.6 m)
Over 500 to 750 kV	35 ft (10.7 m)
Over 750 to 1000 kV	45 ft (13.7 m)

b. While in transit with no load and boom or mast lowered:	
Normal voltage (phase to phase)	Minimum required clearance
to 0.75 kV	4 ft (1.2 m)
Over 0.75 to 50 kV	6 ft (1.3 m)
Over 50 to 345 kV	10 ft (3.5 m)
Over 345 to 700 kV	16 ft (4.9 m)
Over 750 to 1000 kV	20 ft (6.1 m)

Over 1,000 (as established by the utility owner/operator or registered professional engineer who is a qualified person with respect to electrical power transmission and distribution).

STANDARD CRANE HAND SIGNALS

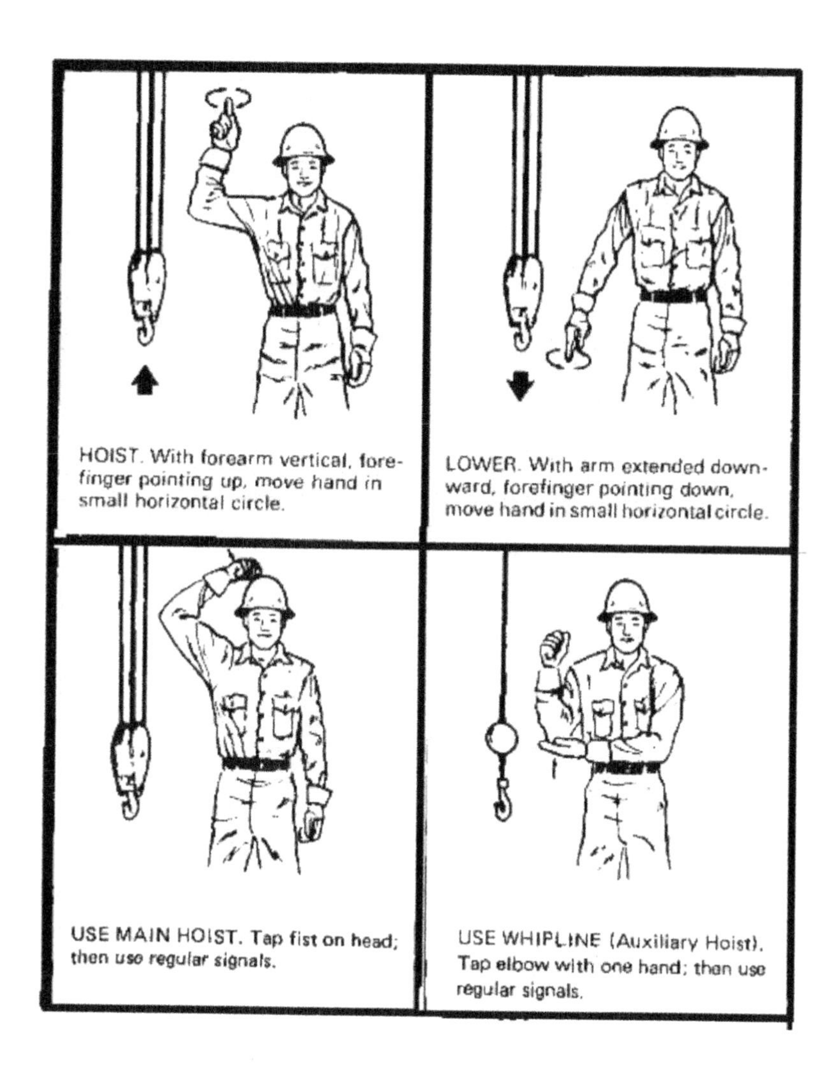

RAISE BOOM. Arm extended, fingers closed, thumb pointing upward.

LOWER BOOM. Arm extended, fingers closed, thumb pointing downward.

MOVE SLOWLY. Use one hand to give any motion signal and place other hand motionless in front of hand giving the motion signal. (Hoist slowly shown as example.)

RAISE THE BOOM AND LOWER THE LOAD. With arm extended, thumb pointing up, flex fingers in and out as long as load movement is desired.

LOWER THE BOOM AND RAISE THE LOAD. With arm extended, thumb pointing down, flex fingers in and out as long as load movement is desired.

SWING. Arm extended, point with finger in direction of swing of boom.

STOP. Arm extended, palm down, move arm back and forth horizontally.

EMERGENCY STOP. Both arms extended, palms down, move arms back and forth horizontally.

EXTEND BOOM (Telescoping Booms). Both fists in front of body with thumbs pointing outward.

RETRACT BOOM (Telescoping Booms). Both fists in front of body with thumbs pointing toward each other.

EXTEND BOOM (Telescoping Boom). One Hand Signal. One fist in front of chest with thumb tapping chest.

RETRACT BOOM (Telescoping Boom). One Hand Signal. One fist in front of chest, thumb pointing outward and heel of fist tapping chest.

OIL FIELD ACRONYMS & ABBREVIATIONS

1P - proven reserves
2D – two-dimensional (**geophysics**)
2P – proved and probable reserves
3C – three components seismic acquisition (x,y and z)
3D – three-dimensional (**geophysics**)
3P – proved, probable and possible reserves
4D – multiple 3Ds overlapping each other (**geophysics**)
7P – prior preparation and precaution prevents piss poor performance, also prior proper planning prevents piss poor performance

A

A - Appraisal (well)
AADE – American Association of Drilling Engineers[1]
AAODC – American Association of Oilwell Drilling Contractors (obsolete; superseded by **IADC**)
AAPG – **American Association of Petroleum Geologists**[2]
AAPL - American Association of Professional Landmen
AAV – Annulus Access Valve
ABAN – Abandonment, (also as AB)
ABSA – Alberta Boilers Safety Association
ACHE – Air-Cooled Heat Exchanger
ACOU – Acoustic
ACQU – Acquisition log
ADEP - Awaiting Development with Exploration Potential, referring to an asset
ADROC: Advanced Rock Properties Report
ADT – Applied Drilling Technology, ADT log
AFE – Authorization For Expenditure, a process of submitting a business proposal to investors
AFP – Active Fire Protection
AGRU – acid gas removal unit
AHBDF – along hole (depth) below **Derrick** floor
AHD – along hole depth
AIRG: Airgun
AIRRE: Airgun Report
AIT: Array Induction Log
AL – appraisal license (United Kingdom), a type of onshore licence issued before 1996
ALARP – as low as reasonably practicable
ALC: Vertical Seismic Profile Acoustic Log Calibration Report
ALR: Acoustic Log Report
aMDEA – activated methyldiethanolamine
AMI - area of mutal interest

AMS – auxiliary measurement service log; Auxiliary Measurement Sonde (temperature)
AMSL – above mean sea level
AMV – annulus master valve[3]
ANACO: Analysis of Core Logs Report
ANARE: Analysis Report
AOF: Absolute Open Flow
AOFP: Absolute Open Flow Potential
APD: Application for Permit to Drill
API – **American Petroleum Institute**: organization which sets unit standards in the oil and gas industry
APPRE – appraisal report
APS: Active Pipe Support
APWD - annular pressure while drilling (tool) **[2]**
ARACL: Array Acoustic Log
ARESV: Analysis of Reservoir
ARI: Azimuthal Resistivity Image
ARRC: Array Acoustic Report
ART – actuator running tool
AS – array sonic processing log
ASCSSV – annulus surface controlled sub-surface valve[3]
ASI – ASI log
ASME – **American Society of Mechanical Engineers**
ASP: Array Sonic Processing
ASTM – American Society for Testing and Materials
ASV – annular safety valve
ATD – application to drill
AUV - authonomus underwater vehicle
AV – annular velocity or apparent viscosity
AVO – **amplitude versus offset** (geophysics)
AWB/V – annulus wing block/valve **(XT)**
AWO – approval for well operation

B
B or b – prefix denoting a number in billions
BA - Bottom Assembly (of a riser)
bbl: barrel
BBSM – **behaviour-based safety** management
Bcf – billion cubic feet (of natural gas)
Bcfe – billion cubic feet (of natural gas equivalent)
BCPD - barrels condensate per day
BDF – below **Derrick** floor
BDL: Bit Data Log
BGL – below ground level (used as a datum for depths in a well)
BGL: Borehole Geometry Log
BGS – **British Geological Survey**
BGT – borehole geometry tool

BGWP: Base of Group Water Protection
BH – bloodhound
BHA – bottom hole assembly (toolstring on **coiled tubing** or **drill pipe**)
BHC – BHC gamma ray log
BHCA – BHC acoustic log
BHCS – BHC sonic log
BHCT – bottomhole circulating temperature
BHL – borehole log
BHP – bottom hole pressure
BHPRP – borehole pressure report
BHSRE – bottom hole sampling report
BHSS – borehole seismic survey
BHT – bottomhole temperature
BHTV – borehole television report
BINXQ – bond index quicklook log
BIOR – biostratigraphic range log
BIORE – biostratigraphy study report
BIVDL – BI/DK/WF/Casing collar locator/gamma ray log
BLI – bottom of logging interval
BO – back-off log
BOB – back on bottom
BOD - Biological Oxygen Demand
boe – barrel(s) of oil equivalent
boed – barrel(s) of oil equivalent per day
BOEM - Bureau Ocean Energy Management
boepd – barrel(s) of oil equivalent per day
BOM – Bill of materials
BOP – bottom of pipe
BOP – **Blowout preventer**
BOPD: Barrels of Oil Per Day
BOREH – borehole seismic analysis
BOREH: Borehole Seismic Analysis
BOSIET – basic offshore safety induction and emergency training
BOTHL – bottom hole locator log
BOTTO – bottom hole pressure/temperature report
bpd – **barrels per day**
BPFL – borehole profile log
BPH – barrels per hour
BPLUG – baker plug
BPV – back pressure valve (goes on the end of coiled tubing a drill pipe tool strings to prevent fluid flow in the wrong direction)
BQL – B/QL log
BRPLG – bridge plug log
BRT – below rotary table (used as a datum for depths in a well)

BS - Bend Stiffener
BS&W – basic sediments and water
BSEE – US: **Bureau of Safety and Environmental Enforcement** (Formerly the MMS)
BSML – below sea mean level
BSR – blind shear rams (**blowout preventer**)
BTEX – benzene, toluene, ethyl-benzene and xylene
BTHL – bottom hole log
BTO/C – break to open/close (valve torque)
BTU – **British thermal units**
BU – bottom up
BUL – bottom up lag
BUR – build-up rate
BVO – ball valve operator
bwd – barrels of water per day (often used in reference to **oil production**)
bwpd – barrels of water per day

C
C&E – well completion and equipment cost
C&S – cased and suspended
C1 – **methane**
C2 – **ethane**
C3 – **propane**
C4 – **butane**
C6 – **hexanes**
C7+ – heavy **hydrocarbon** components
CA – core analysis log
CALI – calliper log
CALOG – circumferential acoustic log
CALVE – calibrated velocity log data
CAPP – **Canadian Association of Petroleum Producers**
CART – cam-actuated running tool (housing running tool)
CART – Cap Replacement Tool
CAS – casing log
CAT - Connector Actuating Tool
CB – core barrel
CBIL – CBIL log
CBL – **cement bond log** (measurement of casing cement integrity)
CBM – choke bridge module – XT choke
CBM – coal-bed methane
CBM – conventional buoy mooring
CCHT – core chart log
CCL – casing collar locator (in perforation or completion operations, the tool provides depths by correlation of the casing string's magnetic anomaly with known casing features)
CCLP – casing collar locator perforation

CCLTP – casing collar locator through tubing plug
CD – core description
CDATA – core data
CDIS – CDI synthetic seismic log
CDP – common depth point (geophysics)
CDP – comprehensive drilling plan
CDP: Common Depth Point
CDRCL – compensated dual resistivity cal. log
CDU – control distribution unit
CE – CE log
CECAN – CEC analysis
CEME – cement evaluation
CERE – cement remedial log
CET – cement evaluation tool
CF – completion fluid
CFD – computational fluid dynamics
CGEL – CG EL log
CGL – core gamma log
CGPH – core graph log
CGR – condensate gas ratio
CGTL – compact **gas to liquids** (production equipment small enough to fit on a ship)
CHCNC – CHCNC gamma ray casing collar locator
CHDTP – calliper HDT playback log
CHECK – checkshot and acoustic calibration report
CHESM – contractor, health, environment and safety management
CHKSR – checkshot survey report
CHKSS – checkshot survey log
CHOPS - Cold Heavy Oil Production with Sand
CHP – casing hanger pressure (pressure in an **annulus** as measured at the **casing hanger**)
CHROM – chromatolog
CHRT – casing hanger running tool
CIBP – cast iron bridge plug
CIDL - chemical injection downhole lower
CIDU - chemical injection downhole upper
CILD – conduction log
CIMV – chemical injection metering valve
CITHP – closed-in tubing head pressure (tubing head pressure when the well is shut in)
CITHP: Closed In Tubing Hanger Pressure
CIV – **chemical injection valve**
CL – core log
CLG – core log and graph
CM – choke module
CMP – common midpoint (geophysics)

CMR – Combinable Magnetic Resonance (**NMR** log tool)
CNCF – field-normalised compensated neutron porosity
CND – compensated neutron density
CNFDP – CNFD true vertical-depth playback log
CNGR – compensated neutron gamma-ray log
CNL – compensated neutron log
CNLFD – CNL/FDC log
CNS – Central **North Sea**
CO – change out (ex. from rod equipment to casing equipment)
COA - Conditions of Approval
COC - Certificate of Conformance
COD - Chemical Oxygen Demand
COL – collar log
COMAN – compositional analysis
COML – compaction log
COMP – composite log
COMPR – completion program report
COMPU – computest report
COMRE – completion record log
CONDE – condensate analysis report
CONDR – continuous directional log
CORAN – core analysis report
CORE – core report
CORG – corgun log
CORIB – coriband log
CORLG – correlation log
COROR – core orientation report
COXY – **carbon/oxygen** log
CP - Cathodic Protection
CPI – CPI log (computer-processed interpretation)
CPI separator – corrugated plate interceptor
CPI: CPI Log
CPICB – CPI coriband log
CPIRE – CPI report
CRA – corrosion-resistant alloy
CRET – cement retainer setting log
CRP – common/central reference point (subsea survey)
CRP – control riser platform
CRT - Casing Running Tool
CRT – Clamp Replacement tool
CsF – Caesium formate (coincidentally also an acronym of the sole large-scale
supplier of caesium formate brine, cabot specialty fluids.)
csg – casing
CSG - Coal Seam Gas
CSHN – cased-hole neutron log

CSI – Combinable Seismic Imager (VSP) log (Schlumberger)
CSMT – core sampler tester log
CSO – complete seal-off
CSPG – **Canadian Society of Petroleum Geologists**
CSR - Corporate Social Responsibility
CST – Chronological Sample Taker log (Schlumberger)
CSTAK – core sample taken log
CSTRE – CST report
CSUG – **Canadian Society for Unconventional Gas**
CT – **coiled tubing**
CTCO – coiled tubing clean-out
CTD – coiled tubing drilling
CTLF – coil tubing lift frame
CTLF – compensated tension lift frame
CTOD – crack tip opening displacement
CTRAC – cement tracer log
CUT – cutter log
CUTTD – cuttings description report
CWOP – complete well on paper
CWOR – completion work over riser
CYBD – cyberbond log
CYBLK – cyberlook log
CYDIP – cyberdip log
CYDN – cyberdon log
CYPRO – cyberproducts log

D

D - Development
D&C – drilling and completions
D&I - direction and inclinitation (MWD borehole deviation survey) **[5]**
DAC – dipole acoustic log
DARCI – Darci log
DAT – wellhead housing drill-ahead tool
DAZD – dip and azimuth display
DBB – double block and bleed
DC – drill centre
DC – drill collar(s)
DCAL – dual caliper log
DCS – Distributed Control System
DD – directional driller or **directional drilling**
DDBHC – DDBHC waveform log
DDET – depth determination log
DDM – Derrick drilling machine (a.k.a. top drive)
DDNL – dual det. neutron life log
DDPT – drill data plot log
DECC – **Department for Energy and Climate Change** (UK)

DECT – decay time
DEFSU – definitive survey report
DELTA – delta-T log
DEN – density log
DEPAN – deposit analysis report
DEPC – depth control log
DESFL – deep induction SFL log
DEV – development well, **Lahee classification**
DEVLG – deviation log
DEXP – D-exponent log
DF – **Derrick** floor
DFIT - Diagnostic Fracture Injection Test
DFR – drilling factual report
DGP – dynamic geohistory plot (3D technique)[4]
DH – drilling history
DHC – depositional history curve
DHPTT – downhole pressure/temperature transducer
DHSV – **downhole safety valve**
DIBHC – DIS BHC log
DIEGR – dielectric gamma ray log
DIF – drill in fluids
DIL – dual-induction log
DILB – dual-induction BHC log
DILL – dual-induction laterolog
DILLS – dual-induction log-LSS
DILSL – dual-induction log-SLS
DIM – directional inertia mechanism
DINT – dip interpretation
DIP – dipmeter lo
DIPAR – dipole acoustic report
DIPBH – dipmeter borehole log
DIPFT – dipmeter fast log
DIPLP – dip lithology pressure log
DIPRE – dipmeter report
DIPRM – dip removal log
DIPSA – dipmeter soda log
DIPSK – dipmeter stick log
DIRS – directional survey log
DIRSU – directional survey report
DIS – DIS-SLS log
DISFL – DISFL DBHC gamma ray log
DISO – dual induction sonic log
DL – development license (United Kingdom), a type of onshore license issued
before 1996
DLIST – dip-list logDIRSU: Directional Survey Report
DIS: DIS-SLS Log
DISFL: DISFL DBHC Gamma Ray Log

DISO: Dual Induction Sonic Log
DLIS
DLIST: Dip List Log
DLL – dual laterolog (deep and shallow resistivity)
DLS – dog-leg severity (directional drilling)
DM - dry mate
DMA – dead-man anchor
DMRP - Density - Magnetic Resonance Porosity (wireline tool)
DNHO – downhole logging
DOA – delegation of authority
DOE – **Department of Energy**, United States
DOWRE – downhole report
DP – **drill pipe**
DP – **dynamic positioning**
DPDV – dynamically positioned drilling vessel
DPL – dual propagation log
DPLD – differential pressure levitated device (or vehicle)
DPRES – dual propagation resistivity log
DPT – deeper pool test, **Lahee classification**
DQLC – dipmeter quality control log
DR – drilling report
DR – dummy-run log
DR: Drilling Report
DRI – drift log
DRLCT – drilling chart
DRLOG – drilling log
DRLPR – drilling proposal/progress report
DRPG – drilling program report
DRPRS – drilling pressure
DRREP – drilling report
DRYRE – drying report
DS – Deviation survey, (also *directional system*)
DSCAN – DSC analysis report
DSI – dipole shear imager
DSL - digital spectralog (Western Atlas)
DSPT – cross-plots log
DST – drill-stem test
DSTG – DSTG log
DSTL – drill-stem test log
DSTND – dual-space thermal neutron density log
DSTPB – drill-stem test true vertical depth playback log
DSTR – drill-stem test report
DSTRE – drill-stem test report
DSTSM – drill-stem test summary report
DSTW – drill-stem test job report/works
DSV – **diving support vessel** or drilling supervisor

DTI - **Department of Trade and Industry** (UK) (obsolete; superseded by **dBERR**, which was then superseded by DECC)
DTPB - CNT true vertical-depth playback log
DTT - depth to time
DWOP - drilling well on paper (a theoretical exercise conducted involving the service-provider managers)
DWQL - dual-water quicklook log
DWSS - dig-well seismic surface log
DXC - DXC pressure pilot report

E
E - Exploration
E&A - Exploration and Appraisal
E&I - Electrical and Instrumentation
E&P - Exploration and Production
EA - Exploration Asset
EAGE - **European Association of Geoscientists & Engineers**[5]
ECD - Equivalent Circulating Density
ECP - External Casing Packer
ECRD - Electrically Controlled Release Device (for abandoning stuck wireline tool from cable)
ECT - External Cantilevered Turret
EDP - Emergency Disconnect Package
EDP - Exploration Drilling Program Report
EDR - Electronic Drilling Recorder
EDR - Exploration Drilling Report
EDS - Emergency Disconnection Sequence
EFL - Electrical Flying Lead
EFR - Engineering Factual Report
EGMBE - Ethylene Glycol Mono-Butyl Ether
EHT - Electric Heat Trace
EHU - Electro-Hydraulic Unit
EL - Electric Log
ELT - Economic Limit Test
EM - EMOP Log
EMG - Equivalent Mud Gradient
EMOP - EMOP Well Site Processing Log
EMP - Electromagnetic Propagation Log
EMS - Environment measurement sonde (wireline multicaliper)
EMW - Equivalent Mud Weight
EN PI - Enhanced Productivity Index Log
ENG - Engineering Log
ENGF - Engineer Factual Report
ENGPD - Engineering Porosity Data
Eni - **Ente Nazionale Idrocarburi** S.p.A. (Italy)[6]
ENJ - Enerjet Log
EOFL - End of Field Life

EOR – Enhanced Oil Recovery
EOT – End of Tubing
EOW – End Of Well Report
EPCM - Engineering Procurement Construction Management
EPCU – Electrical Power Conditioning Unit
EPIDORIS – Exploration and Production Integrated Drilling Operations and Reservoir Information System
EPL – EPL Log
EPLG – Epilog
EPLPC – EPL-PCD-SGR Log
EPS - Early Production System
EPT – Electromagnetic Propagation
EPT: EPT Log EPTNG – EPT-NGT Log
EPU - Electrical Power Unit
EPV - Early Production Vessel
ER(D) – Extended Reach (Drilling)
ERT – Emergency Response Training
ESD – Emergency Shut-Down
ESP – **Electric Submersible Pump**
ETAP – **Eastern Trough Area Project**
ETD - External Turret Disconnectable
ETTD – Electromagnetic Thickness Test
ETU – Electrical Test Unit
EUR – Estimated Ultimate Recovery
EVARE – Evaluation Report
EWR – End Of Well Report
EXL – or XL, Exploration Licence (United Kingdom), a type of onshore licence issued between the First Onshore Licensing Round (1986) and the Sixth (1992)
EZSV – EZSV Log

F

FAC – factual report
FACHV – four-arm calliper log
FANAL – formation analysis sheet log
FAT – factory acceptance testing
FBE – fusion-bonded epoxy
FC – float collar
FCP – final circulating pressure
FCV - Flow Control Valve
FCVE – F-curve log
FDC – formation density log
FDP – field development plan
FDS – functional design specification
FEED – **front-end engineering design**
FER – field equipment room
FEWD – formation evaluation while drilling

FFAC – formation factor log
FFM – full field model
FGEOL – final geological report
FH – full-hole tool joint
FI – final inspection
FI(M) – free issue (materials)
FID – final investment decision
FID - flame ionisation detection
FIL – FIL log
FINST – final stratigraphic report
FINTP – formation interpretation
FIP – flow-induced pulsation
FIT – fairing intervention tool
FIT – fluid identification test
FIT – formation integrity test
FIT – formation interval tester
FIV – flow-induced vibration
FIV – **formation isolation valve**
FL – F log
FLAP - fluid level above pump
FLDF - Flying Lead Deployment Frame
FLIV – flowline injection valve
FLIV – flowline isolation valve
FLNG - Floating liquefied natural gas
FLOG – FLOG PHIX RHGX Log
FLOPR – flow profile report
FLOT – **flying lead orientation tool**
FLOW – flow and buildup test report
FLRA – field-level risk assessment
FLS – fluid sample
FLT – **fault (geology)**
FMEA – failure modes, & effects analysis
FMECA – failure modes, effects, and criticality analysis
FMI – formation micro imaging log (azimuthal microresistivity)
FMP – formation microscan report
FMS – formation multi-scan log; formation micro-scan log
FMTAN – FMT analysis report
FOET – Further Offshore Emergency Training
FOSV – full-opening safety valve
FPDM – fracture potential and domain modelling/mapping[7][8]
FPIT – free-point indicator tool
FPL – flow analysis log
FPLAN – field plan log
FPSO – **floating production storage and offloading vessel**
FPU – floating processing unit
FRA – fracture log
FRARE – fracture report
FRES – final reserve report

FS – fail safe
FSB – flowline support base
FSI – flawless start-up initiative
FSO – floating storage offloading vessel
FT – formation tester log
FTM – fire-team member
FTP – first tranche petroleum
FTRE – formation testing report
FULDI – full diameter study report
FV – funnel viscosity or float valve
FWHP – flowing well-head pressure
FWKO – free water knock-out
FWL – free water level
FWR – final well report
FWV – flow wing valve (also known as production wing valve on a **xmas tree**)

G

G/C – **Gas Condensate**
GAS – Gas Log
GASAN – Gas Analysis Report
GBS – Gravity Based Structure
GBT – Gravity Base Tank
GCLOG – Graphic Core Log
GCT – GCT Log
GDAT – Geodetic Datum
GDE - Gross Depositional Environment
GDIP – Geodip Log
GDT – Gas Down To
GE – Condensate gas equivalent
GE – Ground Elevation (also [GR or GRE])
GEOCH – Geochemical Evaluation
GEODY – GEO DYS Log
GEOEV – Geochemical Evaluation Report
GEOFO – Geological & Formation Evaluation Report
GEOL – Geological Surveillance Log
GEOP – Geophone Data Log
GEOPN – Geological Well Prognosis Report
GEOPR – Geological Operations Progress Report
GEORE – Geological Report
GGRG – Gauge Ring
GIIP – Gas Initially In Place
GIS – **Geographic Information System**
GL – Gas Lift
GL – Ground Level
GLE – Ground Level Elevation (generally in metres above mean sea level)

GLM - Gas Lift Mandrel (alternative name for **Side Pocket Mandrel)**
GLR - Gas Liquid Ratio
GLT - GLT Log
GLV - Gas Lift Valve
GLW -
GOC - Gas Oil Contact
GOM - **Gulf of Mexico**
GOP - Geological Operations Report
GOR - **Gas Oil Ratio**
GOSP - Gas/Oil Separation Plant
GPIT - General Purpose Inclinometry Tool (borehole survey) **[7]**
GPLT - Geol Plot Log
GPM- Gallons Per Mcf
GPSL - Geo Pressure Log
GR - Ground Level (or Elevation //(also [GE or GRE])) or Gamma Ray
GRAD - Gradiometer Log
GRE - Ground Elevation (also [GR or GE])
GRLOG - Grapholog
GRN - Gamma Ray Neutron Log
GRP - Glass Reinforced Plastic
GRSVY - Gradient Survey Log
GRV - Gross Rock Volume
GS - Gas Supplier
GS - Gel Strength
GST - GST Log
GTL - **Gas To Liquid**
GTW - Gas To Wire
GUN - Gun Set Log
GWC - Gas-Water Contact
GWREP - Geo Well Report

H
HAZ - Heat Affected Zone
HAZID - Hazard Identification (meeting)
HAZOP - Hazardous Operation
HC - **Hydrocarbons**
HCAL - HRCC Calliper (in Logs)(in Inches)
HCCS - Horizontal Clamp Connection System
HCM - Horizontal Connection Module. To connect the Xmas Tree to the Manifold
HDA - Helideck Assistant
HDPE - High Density Polyethylene
HDT - High Resolution Dipmeter Log
HDU - Horizontal Drive Unit
HEXT - Hex Diplog

HFE – Human Factors Engineering
HFL – Hydraulic Flying Lead
HGS – High (Specific-)Gravity Solids
HHP – Hydraulic Horsepower
HI - Hydrogen Index
HIPPS – High Integrity Pressure Protection System
HIRA - Hazard Identification and Risk Assessment
HISC – Hydrogen induced stress cracking
HL – Hook Load
HLO – Heavy Load-out (Facility)
HLO – Helicopter Landing Officer
Hmax – Maximum Wave Height
HNGS - flasked Hostile Natural Gamma-ray Spectrometry tool **[8]**
HP – Hydrostatic Pressure
HP: High Pressure
HPGAG – High Pressure Gauge
HPHT – High Pressure High Temperature (same as HTHP)
HPPS – HP Pressure Log
HPU – Hydraulic Power Unit
HPWBM – High Performance Water Base Mud
HRCC – HCAl of Calliper (in Inches)
HRLA - High Resolution Laterolog Array (resistivity logging tool)
Hs – Significant Wave Height
HSE – Health, Safety and Environment or **Health & Safety Executive** (United Kingdom)
HSE: Health & Safety Executive
HSE: Health, Safety and Environment
HTHP – High Temperature High Pressure (same as HPHT)
HTM – Helideck Team Member
HUD – Hold Up Depth
HUET - Helicopter Underwater Escape Training
HVDC – High Voltage Direct Current
HWDP – Heavy Weight Drill Pipe
HWDP – Heavy-Weight Drill Pipe (sometimes spelled Hevi-Wate)
HYPJ – Hyperjet
HYROP – Hydrophone Log

I

I:P – Injector To Producer Ratio
IADC – **International Association of Drilling Contractors**
IAT - Internal Active Turret
IBC – Intermediate Bulk Container
ICD – Inflow Control Device
ICoTA – Intervention and Coiled Tubing Association

ICP – Intermediate Casing Point
ICSS - Integrated Controls and Safety System
ICSU – Integrated Commissioning and Start Up
ICV – Integrated Cement Volume (of Borehole)(in Cubic Metres)
ICV – Interval Control Valve
ID – Inner or Internal Diameter (of a tubular component such as a **casing**)
IDC – Intangible Drilling Costs
IDEL – IDEL Log
IEB – Induction Electro BHC Log
IEL – Induction Electrical Log
IF – Internal Flush tool joint
IH – Gamma Ray Log
IHV – Integrated Hole Volume (of Borehole) (in Cubic Metres)
IJL – Injection Log
IL – Induction Log
ILI – InLine Inspection (Intelligent **Pigging**)
ILOGS – Image Logs
IMAG – Image Analysis Report
IMCA – **International Marine Contractors Association**
IMR - Inspection Maintenance and Repair
INCR – Incline Report
INCRE – Incline Report
INDRS – IND RES Sonic Log
INDT – INDT Log
INDWE: Individual Well Record Report
Information System
INJEC – Injection Falloff Log
INSUR – Inrun Survey Report
INVES – Investigative Program Report
IOC – International Oil Company
IP: Injector to Producer Ratio
IPAA – **Independent Petroleum Association of America**
IPC – Installed Production Capacity
IPLS – IPLS Log
IPR – Inflow Performance Relationship
IPT - Internal Passive Turret
IR – Interpretation Report
IRC – Inspection Release Certificate
IRTJ – IRTJ Gamma Ray Slimhole Log
ISF – ISF Sonic Log
ISFBG – ISF BHC GR Log
ISFCD – ISF Conductivity Log
ISFGR – ISF GR Casing Collar Locator Log
ISFL – ISF-LSS Log
ISFP – ISF Sonic True Vertical Depth Playback Log
ISFPB – ISF True Vertical Depth Playback Log

ISFSL – ISF SLS MSFL Log
ITD - Internal Turret Disconnectable
ITR - Inspection Test Record
ITS – Influx To Surface
IUG - Instrument Utility Gas
IWCF – **International Well Control Federation**
IWOCS – Installation / Workover Control System

J
J&A – Junked and Abandoned
JB – Junk Basket
JIB - Joint Interest Billing
JLT - J-Lay Tower
JT – **Joule-Thomson** (effect/valve/separator)
JU – **Jack-Up drilling rig**
JV - Joint Venture
JVP – Joint Venture Partners/Participants

K
KB – **Kelly** Bushing
KBE – Kelly Bushing Elevation (in meters above sea level, or meters above Ground Level)
KBG – Kelly Bushing Height above Ground Level
KD – Kelly Down
KMW - Kill Mud Weight
KOP – Kick Off Plug
KOP – Kick-Off Point **Directional Drilling**
KRP – Kill Rate Pressure

L
LACT – Lease Automatic Custody Transfer
LAH – Lookahead
LAOT – Linear Activation Override Tool
LARS – Launch & Recovery System
LAS - Log ASCII Standard
LAT – Lowest Astronomical Tide
LCM – Lost Circulation Material
LCNLG – LDT CNL Gamma Ray Log
LD – Lay Down (tubing, rods, etc.)
LDL – Litho Density Log
LDTEP – LDT EPT Gamma Ray Log
LEAKL – Leak Detection Log
LEPRE – Litho-Elastic Property Report
LGR – Liquid Gas Ratio
LGS – Low (specific-)Gravity Solids
LINCO – Liner and Completion Progress Report
LIOG – Lithography Log

LIT – Lead Impression Tool
LITDE – Litho Density Quicklook Log
LITHR – **Lithological** Description Report
LITRE – Lithostratigraphy Report
LITST – Lithostratigraphic Log
LKO – Lowest Known Oil
LL – Laterolog
LMAP – Location Map
LMRP – Lower Marine Riser Package
LMV – Lower Master Valve (on a **Xmas tree**)
LNG – **Liquefied Natural Gas**
LOA – Letter of Authorisation/Agreement/Authority
LOE – Lease Operating Expenses
LOGGN – Logging Whilst Drilling
LOGRS – Log Restoration Report
LOGSM – Log Sample
LOLER – Lifting Operations and Lifting Equipment Regulations
LOT – Leak-Off Test
LOT – Linear Override Tool
LOT – Lock Open Tool
LOT: Leak-Off Test
LOTO – Lock Out / Tag Out
LP – Low Pressure
LPG – **Liquefied Petroleum Gas**
LPH – Litres Per Hour
LPWHH – Low Pressure Well Head Housing
LRP – Lower Riser Package
LSBGR – Long Spacing BHC GR Log
LSD – Land Surface Datum
LSSON – Long Spacing Sonic Log
LT – Linear Time or Lag Time
LTI(FR) – Lost Time Incident (Frequency Rate)
LUMI – Luminescence Log
LVEL – Linear Velocity Log
LVOT – Linear Valve Override Tool
LWD – Logging While Drilling
LWOP – Logging Well on Paper

M

m – metre
M or m – prefix designating a number in thousands (not to be confused with SI prefix M for **mega-** or m for **milli**)
MAASP – Maximum Acceptable [or Allowable] Annular Surface Pressure
MAASP: Maximum Acceptable Annular Shut-in Pressure
MAC – Multipole Acoustic Log
MACL – Multiarm Caliper Log
MAGST – Magnetostratigraphic Report

MAOP – Maximum Allowable Operating Pressure
MARA – Maralog
MAST - sonic tool (records waveform).
MAWP – Maximum Allowable Working Pressure
Mbd – thousand barrels (of oil) per day
Mbod – thousand barrels of oil per day
Mboe – thousand barrels of oil equivalent
Mboed – thousand barrels of oil equivalent per day
Mbpd – thousand barrels of oil per day
MBR - Minimum Bend Radius
MBRO – Multi-Bore Restriction Orifices
MBT – Methylene Blue Test
MBWH - Multi Bowl **Wellhead**
MCD – Mechanical Completion Dossier
Mcf – thousand cubic feet of natural gas
Mcfe – thousand cubic feet of natural gas equivalent
MCHE – Main Cryogenic Heat Exchanger
MCM – Manifold Choke Module
MCS – Manifold & Connection System
MCS – Master Control Station
MD – Measured Depth (see also MDSS)
MD – Measurements/Drilling Log
mD – **millidarcy**, measure of **permeability**, with units of area
MD: Measured Depth
MD: Measurements/Drilling Log
MDEA – **Methyl Diethanolamine** (aMDEA)
MDL – **Methane** Drainage Licence (United Kingdom), a type of onshore licence allowing natural gas to be collected "in the course of operations for making and keeping safe mines whether or not disused"
MDSS – Measured Depth Sub-Sea
MDT – **Modular formation Dynamic Tester**
MEA – **Monoethanolamine**
MEG – **Mono-Ethylene Glycol**
MEIC - Mechanical Electrical Instrumentation Commission
MEPRL – Mechanical Properties Log
MERCR – Mercury Injection Study Report
MERG – Merge FDC/CNL/Gamma Ray/Dual Laterolog/Micro SFL Log
MEST – Micro-electrical scanning tool (a dipmeter, aka. MST or FMS).[9]
MEST: MEST Log
MF – Marsh Funnel (mud **viscosity**)
MFCT – Multifinger Calliper Tool
MGL – Magnelog
MIFR – Mini Frac Log
MINL – Minilog

MIPAL – Micropalaeo Log
MIRU – Move In and Rig Up
MIST – Minimum Industry Safety Training
MIYP – Maximum Internal Yield Pressure
mKB – Meters below **Kelly Bushing**
ML – Microlog, or Mud Log
MLF – Marine Loading Facility
MLH – Mud Liner Hanger
MLL – Microlaterolog
MM or mm – prefix designating a number in millions (or SI unit – millimetre)
MMbd – million barrels per day
mmbd: million barrels per day
MMbod – million barrels of oil per day
mmbod: million barrels of oil per day
MMboe – million barrels of oil equivalent
MMboed – million barrels of oil equivalent per day
MMbpd – million barrels per day
MMcf – million cubic feet (of natural gas)
MMcfe – million cubic feet (of natural gas equivalent)
MMS – **Minerals Management Service**, (United States)
mmscfd – million standard cubic feet per day
mmstb – million stock barrels
MMTPA – Millions of Metric Tonnes per **Annum**
MNP – Merge and Playback Log
MODU – Mobile Offshore Drilling Unit
MOF – Marine Offloading Facility
MOPU – Mobile Offshore Production Unit
MOT – Materials/Marine Offloading Terminal
MPA – Micropalaeo Analysis Report
MPD – Managed Pressure Drilling
MPFM – Multi-Phase Flow Meter
MPK – Merged Playback Log
MPP – Multiphase Pump
MPQT – Manufacturing Procedure Qualification Test
MPS – Manufacturing Procedure Specification
MPV – Multi Purpose Vessel
MQC – Multi Quick Connection Plate
MR – Marine Riser
MR - Morning Report
MRBP - Magna Range Bridge Plug
MRCV – Multi Reverse Circulating Valve
MRIRE – Magnetic Resonance Image Report
MRR - Material Receipt Report
MRX - Magnetic resonance expert (wireline **NMR** tool) **[9]**
MSCT – Mechanical Sidewall Coring Tool **[10]**
MSCT: MSCT Gamma Ray Log

MSFL – Micro SFL Log; Micro-Spherically Focussed Log (resistivity).
MSI – Mechanical and Structural Inspection
MSIP - Modular Sonic Imaging Platform (Sonic Scanner) **[11]**
MSIPC – **Multi Stage inflatable Packer Collar**
MSL – Micro Spherical Log
MSL – **Mean Sea Level**
MSL: Micro Spherical Log
MSS – Magnetic Single Shot
MST – MST EXP **Resistivity Log**
MTBF - Mean Time Between Failure
MTO - Material Take-Off
MTT – MTT Multi-Isotope Trace Tool
MUD – Mud Log
MUDT – Mud Temperature Log
MuSol – Mutual Solvent
MVB – Master Valve Block on Xmas Tree
MW – **Mud weight**
MWD – **Measurement While Drilling**
MWDRE – **Measurement While Drilling** Report
MWS – **Marine Warranty Survey**

N

NACE – **National Association of Corrosion Engineers**
NAPF – Non Aqueous Phase Fluid
NASA – Non Active Side Arm (term used in the **North Sea** for kill wing valve on a **Xmas tree**)
NAVIG – Navigational Log
NB – Nominal Bore
ND - Nipple Down
NDE – Non Destructive Examination
NEUT – Neutron Log
NFG – 'No Fucking Good' used for marking damaged equipment, not to be confused with NG being natural gas
NFI – No Further Investment
NFW – New Field Wildcat, **Lahee classification**
NG – Natural Gas
NGDC – National Geoscience Data Centre (United Kingdom)
NGL – Natural Gas Liquids
NGLQT – NGT QL Log
NGR – Natural Gamma Ray
NGRC – National Geological Records Centre (United Kingdom)
NGS – NGS Log
NGSS – NGS Spectro Log
NGT – Natural Gamma Ray Tool
NGT: NGT Log
NGTLD – NGT LDT QL Log

NGTR – NGT Ratio Log
NHDA – National Hydrocarbons Data Archive (United Kingdom)
NHPV – Net **Hydrocarbon** Pore Volume
NMDC - Non Magnetic Drill Collar
NMHC – Non-Methane Hydrocarbons
NMR – Nuclear Magnetic Resonance **Log**
NMVOC – Non-Methane Volatile Organic Compounds
NNS – Northern **North Sea**
NOC – **National Oil Company**
NOISL – Noise Log
NORM – Naturally Occurring Radioactive Material
NP – Non Producing well (as opposed to P – Producing well)
NPD – **Norwegian Petroleum Directorate**
NPS – Nominal Pipe Size (sometimes NS)
NPSH – Net Positive Suction Head
NPV – Net Present Value
NPW – New Pool Wildcat, **Lahee classification**
NRI - Net Revenue Interest
NRV – Non Return Valve
NS – **North Sea**; can also refer to the **North Slope Borough, Alaska**, the **North Slope**, which includes **Prudhoe Bay Oil Field** (the largest US oil field), **Kuparuk Oil Field**, Milne Point, Lisburne, and Point McIntyre among others.
NTHF – Non-Toxic High Flash
NTU – Nephelometric Turbidity Unit
NUMAR – **Magnetic Resonance** Image Log

O

O&G – Oil and Gas
O&M – Operations and Maintenance
OBCS – Ocean Bottom Cable System
OBDTL – OBDT Log
OBEVA – OBDT Evaluation Report
OBM – Oil-Based Mud
OCI - Oil Corrosion Inhibitor (Vessels)
OCIMF – **Oil Companies International Marine Forum**
OCL – Quality Control Log
OCTG – Oil Country Tubular Goods (oil well casing, tubing, and drill pipe)[10]
OD – Outer Diameter (of a tubular component such as **casing**)
ODT – Oil Down To
OEM – Original Equipment Manufacturer
OFST – Offset Vertical Seismic Profile
OH – Openhole
OH – Openhole Log
OHC – Open Hole Completion
OI - Oxygen Index
OIM – **Offshore Installation Manager**

OMRL – Oriented Micro-Resistivity **Log**
ONAN – Oil Natural Air Natural cooled transformer
ONNR - Office of Natural Resources Revenue (formerly MMS)
OOE – Offshore Operation Engineer (senior technical authority on an offshore **oil platform**)
OOIP – Original Oil In Place
OOT – Out of Tolerance
OPEC – **Organization of Petroleum Exporting Countries**
OPITO - Offshore Petroleum Industry Training Organization
OPL – Operations Log
OPRES – Overpressure Log
ORICO – Oriented Core Data Report
OS&D - Over, Short & Damage (Report)
OT – A Well On Test
OT – Off Tree
OTDR – Optical Time Domain Reflectometry
OTIP – Operational Testing Implementation Plan
OTL – Operations Team Leader
OTSG - One-time Through Steam Generator
OUT – Oil Up To
OUT – Outpost, **Lahee classification**
OVCH – Oversize Charts
OVID – Offshore Vessel Inspection Database
OWC – Oil Water Contact
OWC – Oil-Water Contact

P

P – Producing well (as opposed to NP – Non Producing)
P&A – plugged and abandoned (of a well)
PA - Polyamide
PA - Producing Asset with Exploration Potential
PADPRT – Pressure Assisted Drillpipe Running Tool
PAGA – Public Address General Alarm
PAL – Palaeo Chart
PALYN – **Palynological** Analysis Report
PAR – Pre-Assembled Rack
PAU – Pre-Assembled Unit
PBD – Pason Billing System
PBDMS – Playback DMSLS Log
PBHL - Proposed Bottom Hole Location
PBR – Polished Bore Receptacle (component of a **completion string**)
PBTD – Plug Back Total Depth
PBU – Pressure Build Up (applies to integrity testing on valves)
PCA - Production Concession Agreement
PCB – **Poly Chlorinated BiPhenyl**
PCCL – Perforation Casing Collar Locator Log

PCDM – Power and Control Distribution Module
PCKR – Packer
PCOLL – Perforation and Collar
PDC – Perforation Depth Control
PDC – Polycrystalline Diamond Compact (a type of drilling bit)
PDC: Perforation Depth Control
PDC: Polycrystalline Diamond Composite
PDG/PDHG – Permanent Downhole Gauge
PDKL – PDK Log
PDKR – PDK 100 Report
PDM - **Positive Displacement Motor**
PDNP – Proved Developed Not Producing
PDP – Proved Developed Reserves
PDPM – Power Distribution Protection Module
PE – Petroleum Engineer
PE - Polyethylene
PE – Professional Engineer
PEA – Palaeo Environment Study Report
PEA: Palaeo Environment Study Report
PEDL – Petroleum Exploration and Development Licence (United Kingdom), introduced in the 8th Onshore Licensing Round
PENL – Penetration Log
PEP – PEP Log
PERDC – Perforation Depth Control
PERFO – Perforation Log
PERM – Permeability
PERML – Permeability Log
PESBG: Petroleum Exploration Society of Great Britain
PETA – Petrographical Analysis Report
PETD – Petrographic Data Log
PETLG – Petrophysical Evaluation Log
PETPM – **Petrography** Permeametry Report
PETRP – Petrophysical Evaluation Report
PEX - Platform Express toolstring (resistivity, porosity, imaging)
PFC – PFC Log
PFE – Plate/Frame Heat Exchanger
PFHE – Plate Fin Heat Exchanger
PFPG – Perforation Plug Log
PFREC – Perforation Record Log
PG – Pressure Gauge (Report)
PGB – Permanent Guide Base
PGOR – Produced Gas Oil Ratio
PH – Phasor Log
PHASE – Phasor Processing Log
PHB - Pre-Hydrated Bentonite
PHC - Passive Heave Compensator

PHOL – Photon Log
PHPU - Platform Hydraulic Power Unit
PHYFM – Physical Formation Log
PI – Productivity Index or (Permit Issued)
PINTL – Production Interpretation
PIP – Pipe in Pipe
PL – Production Licence
PLEM – Pipeline End Manifold
PLES -Pipeline End Structure
PLET – Pipeline End Termination
PLG – Plug Log
PLS – Position Location System
PLSV - Pipelay Support Vessel
PLT – **Production Logging Tool**
PLTQ – **Production Logging Tool** Quicklook Log
PLTRE – **Production Logging Tool** Report
PMI – Positive Material Identification
PMV -Production Master Valve
PNP - Proved Not Producing
POB – Personnel on Board
POBM – Pseudo-Oil-Based Mud
POF – Permanent Operations Facility
PON – Petroleum Operations Notice (United Kingdom)
POOH / POH – Pull Out Of Hole
POR – Density Porosity Log
POSFR – Post Fracture Report
POSTW – Post Well Appraisal Report
POSWE – Post Well Summary Report
PP – DXC Pressure Plot Log
PP – Pump Pressure
PP: Pressure Plot Log
PPC - Powered Positioning Caliper (Schlumberger dual-axis wireline caliper tool) **[12]**
ppcf – Pounds Per Cubic Foot
PPE - Personal Protective Equipment
PPE – Preferred Pressure End
ppg – Pounds Per Gallon
PPI – Post Pipelay Installation
PPI – Post Production Inspection/Intervention
PPI: Post Pipelay Installation
PPS – Production Packer Setting
pptf – Pounds (per square inch) Per Thousand Feet (of depth) – a unit of fluid density/pressure
PQR - Procedure Qualification Record
PR2 – Testing regime to API6A Annex F
PRA – Production Reporting & **Allocation**
PREC – Perforation Record

PRESS – Pressure Report
PROD – Production Log
PROTE – Production Test Report
PROX – Proximity Log
PRSRE – Pressure Gauge Report
PSA – Production Service Agreement
PSA – Production Sharing Agreement
PSANA – Pressure Analysis
PSC – Production Sharing Contract
PSD – Planned Shut-Down
PSIA – Pounds Per Square Inch Atmospheric
PSIG – Pounds Per Square Inch Gauge
PSL – Product Specification Level
PSLOG – Pressure Log
PSP – Pseudostatic Spontaneous Potential
PSPL – PSP Leak Detection Log
PSQ – Plug Squeeze Log
PST – PST Log
PSV – Pipe Supply Vessel
PSV – Pressure Safety Valve
PSVAL – Pressure Evaluation Log
PTA/S – Pipeline Termination Assembly/Structure
PTO - Permit to Operate
PTSET – Production Test Setter
PTTC – Petroleum Technology Transfer Council, United States
PTW – Permit to Work
PU – Pick Up (tubing, rods, power swivel, etc.)
PUD – Proved Undeveloped Reserves
PUN – Puncher Log
PUN: Puncher Log
PUR – Plant Upset Report
PUWER – **Provision and Use of Work Equipment Regulations**
PV – Plastic Viscosity
PVDF - Polyvinylidene Fluoride
PVT – Pressure Volume Temperature
PVTRE – Pressure Volume Temperature Report
PW – Produced Water
PWB – Production Wing Block (XT)
PWHT – Post Weld Heat Treat
PWRI – Produced Water Re-Injection
PWV – Production Wing Valve (also known as a flow wing valve on a **Xmas tree**)

Q

QC – Quality control
QCR – Quality Control Report
QL – Quicklook Log

R

RAC – Ratio Curves
RACI – Responsible / Accountable / Consulted / Informed
RAM – Reliability, Availability, and Maintainability
RAWS – Raw Stacks VSP Log
RBP – Retrievable Bridge Plug
RCA – **Root Cause Analysis**
RCD – Rotating Control Device
RCI – Reservoir Characterization Instrument (for downhole fluid measurements e.g. spectrometry, density, etc) **[13]**
RCKST – Rig Checkshot
RCL – Retainer Correlation Log
RCM – Reliability Centred Maintenance
RCR – Remote Component Replacement (Tool)
RCU – Remote Control Unit
RDMO - Rig Down Move Out
RE – Reservoir Engineer
REOR – Reorientation Log
RE-PE – Re-Perforation Report
RESAN – Reservoir Analysis
RESDV - Riser Emergency Shut Down Valve
RESEV – Reservoir Evaluation
RESFL – Reservoir Fluid
RESI – Resistivity Log
RESL – Reservoir Log
RESOI – **Residual Oil**
REZ – Renewable Energy Zone (United Kingdom)
RF – Recovery Factor
RFLNG – Ready for Liquefied Natural Gas
RFMTS – **Repeat Formation Tester**
RFSU – Ready For Start-Up
RFT – **Repeat Formation Tester**
RFTRE – **Repeat Formation Tester** Report
RFTS – **Repeat Formation Tester** Sample
RIGMO – Rig Move
RIH – Run In Hole
RIMS – Riser Integrity Monitoring System
RITT – Riser Insertion Tube (Tool)
RKB – Rotary Kelly Bushing (a datum for measuring depth in an oil well)
RKB – Rotary Kelly Bushings
RLOF – Rock Load-out Facility
RMLC – Request for Mineral Land Clearance
RMP – Reservoir Management Plan
RMS – Ratcheting Mule Shoe
RMS – Riser Monitoring System
RMS: Ratcheting Mule Shoe

RNT – RNT Log
ROCT – Rotary Coring Tool
ROP – Rate of Penetration
ROP - Rate of Perforation
ROT – Remote Operated Tool
ROV/WROV – Remotely Operated Vehicle/WorkClass Remotely Operated Vehicle, used for **subsea** construction and maintenance
ROV: Remotely Operated Vehicle
ROWS – Remote Operator Workstation
ROZ – Recoverable Oil Zone
RPCM – Ring Pair **Corrosion Monitoring**
RPM – Revolutions Per Minute, (Rotations Per Minute)
RROCK – Routine Rock Properties Report
RSS – Rig Site Survey
RSS – Rotary Steerable Systems
RSS: Rig Site Survey
RST – Reservoir Saturation Tool (Schlumberger) Log
RST: RST Log
RTTS – Retrievable Test-Treat-Squeeze (packer)
RWD – Reaming While Drilling
RWD: RWD Log

S
SABA – Supplied Air Breathing Apparatus
SAGD – Steam Assisted Gravity Drainage
SALM – Single Anchor Loading Mooring
SAM – Subsea Accumulator Module
SAML – Sample Log
SAMTK – Sample Taker Log
SANDA – Sandstone Analysis Log
SAT – SAT Log
SAT – Site Acceptance Test
SAT: SAT Log
SB – SIT-BO Log
SBF – Synthetic Base Fluid
SBM – Synthetic Base Mud
SBT – Segmented Bond Tool
SC – Seismic Calibration
SCADA – Supervisory Control and Data Acquisition
SCAL – **Special core analysis**
SCAP – Scallops Log
SCBA – Self Contained Breathing Apparatus
SCDES – Sidewall Core Description
scf – standard cubic feet (of natural gas)
SCHLL – **Schlumberger** Log also SCHLO, SCHLU
SCM(MB) – Subsea Control Module (Mounting Base)
SCO – **Synthetic crude oil**

SCRS – Sidewall Cores
SCSSSV: Surface Controlled Sub Surface Safety Valve
SD – Sonic Density
SDFD – Shut Down For Day
SDFN – Shut Down For Night
SDIC – Sonic Dual Induction
SDL – Supplier Document List
SDM/U – Subsea Distribution Module/Unit
SDPBH – SDP Bottom Hole Pressure Report
SDT – Step Draw-down Test (sometimes SDDT)
SDU/M – Subsea Distribution Unit/Module
SEA – Strategic Environmental Assessment (United Kingdom)
SECGU – Section Gauge Log
SEDHI – Sedimentary History
SEDIM – Sedimentology
SEDL – Sedimentology Log
SEDRE – Sedimentology Report
SEM – Subsea Electronics Module
Semi (or Semi-Sub) – Semi-Submersible Drilling Rig
SEPAR – Separator Sampling Report
SEQSU – Sequential Survey
SFERAE – Global Association for the use of knowledge on Fractured Rock in a state of Stress, in the field of Energy, Culture and Environment[11]
SFL – Steel Flying Lead
SG – Static Gradient
SGR - Shale Gouge Ratio
SGSI – Shell Global Solutions International
SGUN – Squeeze Gun
SHA – Sensor Harness Assembly
SHDT – Stratigraphic High Resolution Dipmeter Tool
SHDT: SHDT Log
SHO – Stab and Hinge Over
SHOCK – Shock Log
SHOWL – Show Log
SI – Shut In well
SI/TA – Shut In/Temporarily Abandoned
SICP – Shut-In Casing Pressure
SIDPP – Shut-In Drill Pipe Pressure
SIDSM – Sidewall Sample
SIGTTO – Society of International Gas Tanker and Terminal Operators
SIMCON – Simultaneous Construction
SIMOPS – Simultaneous Operations
SIP – Shut In Pressure
SIPES – Society of Independent Professional Earth Scientists, United States

SIT - (Casing) Shoe Integrity Test
SIT – **System Integration Test** FR SIT – Field Representation SIT
SITHP – Shut In Tubing Hanger/Head Pressure (another term for CITHP)
SITT: Single TT Log
SIWHP – Shut-in Well Head Pressure
SKPLT – Stick Plot Log
SL – Seismic Lines
SLS – SLS GR Log
SLT – SLT GR Log
SMA – Small Amount
SMLS – Seamless PipeMPP:
SMO – Suction Module
SMPC- Subsea Multiphase Pump: Pump, which can increase flowrate and pressure of the untreated wellstream
SN – Seat Nipple
SNAM – **Societá NAzionale Metanodotti** now Snam S.p.A. (Italy)[12]
SNP – Sidewall Neutron Porosity
SNS – Southern **North Sea**
SOBM – Synthetic Oil Based Mud
SOLAS – Safety of Life at Sea
SONCB – Sonic Calibration Log
SONRE – Sonic Calibration Report
SONWR – Sonic Waveform Report
SONWV – Sonic Waveform Log
SOW _ Slip-On Wellhead
SP - Set Point
SP – Spontaneous Potential (Well Log)
SP – **Shot point** (geophysics)
SPCAN – **Special core analysis**
SPCU - Subsea Power and Control Unit
SPE – **Society of Petroleum Engineers**[13]
SPEAN – Spectral Analysis
SPEL – Spectralog
SPFM – Single Phase Flow Meter
SPH – SPH Log
SPM – **Side Pocket Mandrel** or Strokes Per Minute (of a positive-displacement pump)
SPMT – **Self-Propelled Modular Transporter**
SPOP – Spontaneous Potential Log
SPP – Stand Pipe Pressure
SPR: Side Pocket Mandrel
SPR: Strokes Per Minute
SPROF – Seismic Profile
SPS – Subsea Production Systems
SPT – Shallower Pool Test, **Lahee classification**

SPWLA - **Society of Petrophysicists and Well Log Analysts**[14]
SQL – Seismic Quicklook Log
SR – Shear Rate
SRD – Seismic Reference Datum, an imaginary horizontal surface at which TWT is assumed to be zero
SREC – Seismic Record Log
SRT – Site Receival Test
SS – Subsea, as in a datum of depth, e.g. TVDSS (True Vertical Depth Subsea)
SS: Subsea
SSCC – **Sulphide Stress Corrosion Cracking**
SSCP -Subsea Cryogenic Pipeline
SSCS – Subsea Control System
SSD – Sliding Sleeve Door
SSD – Sub Sea Level depth (in Metres or Feet, positive value in downwards direction with respect to the Geoid)
SSG – Sidewall Sample Gun
SSIV – Subsea Isolation Valve
SSM – Subsea Manifolds
SSMAR – Synthetic Seismic Marine Log
SSPLR – **Subsea Pig Launcher/Receiver**
SSSL – Supplementary Seismic Survey Licence (United Kingdom), a type of onshore licence
SSSV – **Sub-Surface Safety Valve**
SSTT – **Subsea Test Tree**
SSTV - Subsea Test Valve
SSU – Subsea Umbilicals
SSV – Surface Safety Valve
SSWI – Subsea Well Intervention
STAB – Stabiliser
STAGR – Static Gradient Survey Report
STB – stock tank barrel
STC – STC Log
STFL – Steel Tube Fly Lead
STGL – Stratigraphic Log
STIMU – Stimulation Report
STKPT – Stuck Point
STL – STL Gamma Ray Log
STL - Submerged Turret Loading
STL: STL Gamma Ray Log
STOIIP – Stock Tank Oil Initially In Place
STOP – Safety Training Observation Program
STP - Submerged Turret Production
STRAT – **Stratigraphy**, stratigraphic
STRRE – **Stratigraphy** Report
STSH – String Shot

STU - Steel Tube Umbilical
SUML – Summarised Log
SUMRE – Summary Report
SUMST – Geological Summary Sheet
SURF – Subsea/Umbilicals/Risers/Flowlines
SURFR – Surface Sampling Report
SURRE – Survey Report
SURVL – Survey Chart Log
SUT(A/B) – Subsea Umbilical Termination (Assembly/Box)
SUTU – Subsea Umbilical Termination Unit
SV – Sleeve Valve or Standing Valve
SW – Salt Water
SWD – Salt Water Disposal Well
SWE – Senior Well Engineer
SWHE – Spiral Wound Heat Exchanger
SWOT – Strengths, Weaknesses, Opportunities and Threats
SYNRE – Synthetic Seismic Report
SYSEI – Synthetic Seismogram Log

T

T – well flowing to Tank
TA – Temporarily Abandoned well
TA – Top Assembly
TAC – Tubing Anchor
TAGOGR – Thermally Assisted Gas/Oil Gravity Drainage
TAN - Total Acid Number
TAPLI – Tape Listing
TAPVE – Tape Verification
TAR – True Amplitude Recovery
TB – Tubing Puncher Log
TBT – Through Bore Tree
TCA – Total Corrosion Allowance
TCF – Temporary Construction Facilities
TCF – Trillion Cubic Feet(of gas)
TCI – Tunsgten Carbide Insert (a type of rollercone drillbit)
TCPD – Tubing-Conveyed Perforating Depth
TCU- Thermal Combustion Unit
TD – Target Depth
TD – Total Depth (depth of the end of the well; also a verb, to reach the final depth, used as an acronym in this case)
TD: Total depth
TDD – Total Depth (Driller)
TDL – Total Depth (Logger)
TDM – Touch Down Monitoring
TDP - Touch Down Point
TDS – Top Drive System
TDS – Total Dissolved Solids
TDS: Top Drive System

TDT – TDT Log
TDT GR – TDT Gamma Ray Casing Collar Locator Log
TDT: TDT Log
TDTCP – TDT CPI Log
TEFC – Totally Enclosed Fan Cooled
TEG - Thermal Electric Generator
TEG – Tri-Ethylene Glycol
TELER – Teledrift Report
TEMP- Temperature Log
TFE – TotalFinaElf (obsolete; Now **Total S.A.**) major French multinational oil company[15]
TFL – Through Flow Line
TFM – **TaskForceMajella Research Project**
TGB – Temporary Guide Base
TGOR – Total Gas Oil Ratio (GOR uncorrected for **gas lift** gas present in the **production fluid**)
TH – **Tubing Hanger**
THERM – Thermometer Log
THP – Tubing Hanger Pressure (pressure in the **production tubing** as measured at the **tubing hanger**)
THRT – Tubing Hanger Running Tool
TIE – Tie In Log
TIEBK – Tieback Report
TIH – Trip Into Hole
TIW – Texas Iron Works (pressure valve)
TLI – Top of Logging Interval
TLOG – Technical Log
TMCM – Tranverse Mercator Central Meridian
TMD – Total Measured Depth in a wellbore
TNDT – Thermal Neutron Decay Time
TNDTG – Thermal Neutron Decay Time/Gamma Ray Log
TOC – Top Of Cement
TOC – **Total Organic Content/Carbon**
TOC: Top Of Cement
TOFD – Time of first data sample (on seismic trace)
TOFS – Time of first surface sample (on seismic trace)
TOH – Trip Out of Hole
TOL – Top Of Liner
TOOH – Trip Out Of Hole
TORAN – Torque and Drag Analysis
TPERF – Tool Performance
TQM – Total Quality Management
TR – Temporary Refuge
TRA – Top Riser Assembly
TRA – Tracer Log
TRACL – Tractor Log
TRCFR – Total Recordable Case Frequency Rate

TRD – Total Report Data
TREAT – Treatment Report
TREP – Test Report
TRIP – Trip Condition Log
TRSCSSSV – **Tubing Retrievable Surface Controlled Sub-Surface Safety Valve**
TRSV – **Tubing Retrievable Safety Valve**
TRT – Tree Running Tool
TSA – Thermally Sprayed Aluminium
TSI – Temporarily Shut In
TT – Torque Tool
TT – Transit Time Log
TT: transit time
TTRD – Through Tubing Rotary Drilling
TUM – Tracked Umbilical Machine
TUTB – Topside Umbilical Termination Box
TV/BIP – Ratio of Total Volume (ore and overburden) to Bitumen In Place
TVBDF – True Vertical Depth Below Derrick Floor
TVD – True Vertical Depth
TVDPB – True Vertical Depth Playback Log
TVDSS – True Vertical Depth Sub Sea
TVELD – Time and Velocity to Depth
TVRF – True Vertical Depth versus **Repeat Formation Tester**
TWT – Two-Way Time (seismic)
TWTTL – Two-Way Travel Time Log

U
UBHO - Universal Bottom Hole Orientation (sub)
UBI – Ultrasonic Borehole Imager
UBIRE – Ultrasonic Borehole Imager Report
UCH - Umbilical Connection Housing
UCSU – Upstream Commissioning and Start-Up
UFJ – Upper Flex Joint
UGF – Universal Guide Frame
UIC – Underground Injection Control
UKCS – United Kingdom **Continental Shelf**
UKOOA – **United Kingdom Offshore Operators Association**
UKOOG – **United Kingdom Onshore Operators Group**
ULCGR – Uncompressed LDC CNL Gamma Ray Log
UMCA – Umbilical Midline Connection Assembly
UMV – Upper Master Valve (from a **Xmas tree**)
UPB - Unmanned Production Buoy
UPR – Upper Pipe Ram
URT – Universal Running Tool
USGS – **United States Geological Survey**
USIT - Ultrasonic imaging tool (cement bond logging) **[14]**
UTA/B – Umbilical Termination Assembly/Box

UTA: Umbilical Termination Assembly
UTM – **Universal Transverse Mercator**
UWI – Unique Well Identifier
UWILD – Underwater Inspection in Lieu of Dry Docking

V
VDENL – Variation Density Log
VDL – VDL Log (Variable Density Log)
VDU – Vacuum Distillation Unit, used in processing **bitumen**
VELL – Velocity Log
VERAN – Verticality Analysis

VERIF – Verification List
VERLI – Verification Listing
VERTK – Vertical Thickness
VIR – Value Investment Ratio
VISME – Viscosity Measurement
VISME: Viscosity Measurement
VIV – Vortex induced Vibration
VIV: Vortex induced Vibration
VLP – Vertical Lift Performance
VLS - Vertical Lay System
VLTCS – Very Low Temperature Carbon Steel
VO – Variation Order
VOCs – Volatile Organic Compounds
VOR – Variation Order Request
VRR – Voidage Replacement Ratio
VS – Vertical Section
VSD – Variable Speed Drive
VSI - Versatile Seismic Imager (Schlumberger VSP tool)[16]
VSP – Vertical Seismic Profile
VSPRO – Vertical Seismic Profile
VTDLL – Vertical Thickness Dual Laterolog
VTFDC – Vertical Thickness FDC CNL Log
VTISF – Vertical Thickness ISF Log
VWL – Velocity Well Log
VXT – Vertical Christmas Tree

W
W – Watt
W/C – Water Cushion
WABAN – Well Abandonment Report
WAG – Water Alternating Gas (describes an injection well which alternates between water and gas injection)
WALKS – Walkaway Seismic Profile
WATAN – Water Analysis
WAV3 – Amplitude (in Seismics)

WAV4 – Two Way Travel Time (in Seismics)
WAV5 – Compensate Amplitudes
WAVF – Waveform Log
WBM – Water-Based drilling Mud
WC – Watercut / Wildcat well
WCC - Work Control Certificate
WE – Well Engineer
WEG – **Wireline Entry Guide**
WELDA – Well Data Report
WELP – Well Log Plot
WEQL – Well Equipment Layout
WESTR – Well Status Record
WESUR – Well Summary Report
WF – Water Flooding
WFAC – Waveform Acoustic Log
WGEO – Well Geophone Report
WGFM – Wet Gas Flow Meter
WGR - Water Gas Ratio
WGUNT – Water Gun Test
WH – Well History
Wh – White
WH: Well History
WHIG – Whitehouse Gauge
WHM – Wellhead Maintenance
WHMIS – WorkPlace Hazardous Material Information Systems
WHP – Wellhead Pressure
WHP – Wellhead Pressure
WHSIP - Wellhead Shut-In Pressure
WI – Water Injection
WI – Working Interest
WI: Water Injection
WI: Working Interest
WIPSP – WIP Stock Packer
WIT- Water Investigation Tool
WITS – Well Site Information Transfer System
WLC – Wireline Composite Log
WLL – **Wireline Logging**
WLSUM – Well Summary
WLTS – Well Log Transaction System
WLTS Well Log Tracking System
WM - Wet Mate
WO – Well in Work Over
WO/O – Waiting on Orders
WOA – Well Operations Authorization
WOB – Weight On Bit
WOC – Water Oil Contact (or Oil Water Contact)
WOC: Wait on Cement

WOE – Well Operations Engineer (a key person of **Well services**)
WOM – Wait(ing) on Material
WOR – Water-Oil Ratio
WORKO – Workover
WOS – West of **Shetland**, oil province on the **UKCS**
WOW – Wait(ing) On Weather
WP – Well Proposal or Working Pressure
WPC – Water Pollution Control
WPLAN – Well Course Plan
Plan
WPQ/S – Weld Procedure Qualification/Specification
WPR – Well Prognosis Report
WQ – A textural parameter used for CBVWE computations (**Halliburton**)
WQCA – Water Quality Control Act
WQCB – Water Quality Control Board
WR – Wet Resistivity
WR – Wireline Retrievable (as in a WR Plug)
WRS – Well Report Sepia
WRSCSSV – **Wireline Retrievable Surface Controlled Sub-Surface Valve**
WSCL – Well Site Core Log
WSE – Well Seismic Edit
WSERE – Well Seismic Edit Report
WSHT – Well Shoot
WSL – Well Site Log
WSO – Water Shut Off
WSOG – Well Specific Operation Guidelines
WSP – Well Seismic Profile
WSR – Well Shoot Report
WSS – Well Services Supervisor (leader of **Well services** at the wellsite)
WSS – Working Spread Sheet (for logging)
WSSAM – Well Site Sample
WSSOF – WSS Offset Profile
WSSUR – Well Seismic Survey Plot
WSSVP – WSS VSP Raw Shots
WSSVS – WSS VSP Stacks
WST - Well Seismic Tool (checkshot)
WSTL – Well Site Test Log
WSU – Well Service Unit
wt - wall thickness
WT – Well Test
WTI – **West Texas Intermediate** benchmark crude
WUT – Water Up To
WVS – Well Velocity Survey

X

XC – Cross-connection, cross correlation
XL – or EXL, Exploration Licence (United Kingdom), a type of onshore licence issued between the First Onshore Licensing Round (1986) and the Sixth (1992)
XL: Exploration Licence
Xln – Crystalline (minerals)
XLPE -Cross-linked Polyethylene
XMAC – Cross-Multipole Array Acoustic log
XMAC-E – XMAC Elite (Next generation of XMAC)
XMRI – Extended Range Micro Imager (**Halliburton**)
XMT/XT/HXT – **XMas Tree** (Christmas Tree, the valve assembly on a production well-head)
XO – Cross-Over
XOM – Exxon Mobil
XOV – Cross-Over Valve
XPERM – Matrix Permeability in the X-Direction
XPHLOC – Crossplot Selection for XPHI
XPOR – Crossplot Porosity
XPT – Formation Pressure Test log (**Schlumberger**)
XYC – XY Calliper Log (**Halliburton**)

Y

yd – yard
yl – Holdup Factor
YP – Yield Point
yr – year

Z

Z – Depth, in the geosciences referring to the depth dimension in any x,y,z data.
ZDENP – Density Log
ZDL – Compensated Z-Densilog
ZOI – Zone of Influence

See more at: http://www.allacronyms.com

PROCESSING GLOSSARY

ABSORPTION The disappearance of one substance into another so that the absorbed substance loses its identifying characteristics, while the absorbing substance retains most of its original physical aspects. Used in refining to selectively remove specific components from process streams.
ACID TREATMENT A process in which unfinished petroleum products such as gasoline, kerosene, and lubricating oil stocks are treated with sulfuric acid to improve color, odor, and other properties.
ADDITIVE Chemicals added to petroleum products in small amounts to improve quality or add special characteristics.
ADSORPTION Adhesion of the molecules of gases or liquids to the surface of solid materials.
AIR FIN COOLERS A radiator-like device used to cool or condense hot hydrocarbons; also called fin fans.
ALICYCLIC HYDROCARBONS Cyclic (ringed) hydrocarbons in which the rings are made up only of carbon atoms.
ALIPHATIC HYDROCARBONS Hydrocarbons characterized by open-chain structures: ethane, butane, butene, acetylene, etc.
ALKYLATION A process using sulfuric or hydrofluoric acid as a catalyst to combine olefins (usually butylene) and isobutane to produce a high-octane product known as alkylate.
API GRAVITY An arbitrary scale expressing the density of petroleum products.
AROMATIC Organic compounds with one or more benzene rings.
ASPHALTENES The asphalt compounds soluble in carbon disulfide but insoluble in paraffin naphthas.
ATMOSPHERIC TOWER A distillation unit operated at atmospheric pressure.
BENZENE An unsaturated, six-carbon ring, basic aromatic compound.
BLEEDER VALVE A small-flow valve connected to a fluid process vessel or line for the purpose of bleeding off small quantities of contained fluid. It is installed with a block valve to determine if the block valve is closed tightly.
BLENDING The process of mixing two or more petroleum products with different properties to produce a finished product with desired characteristics.
BLOCK VALVE A valve used to isolate equipment.
BLOWDOWN The removal of hydrocarbons from a process unit, vessel, or line on a scheduled or emergency basis by the use of pressure through special piping and drums provided for this purpose.

BLOWER Equipment for moving large volumes of gas against low-pressure heads.
BOILING RANGE The range of temperature (usually at atmospheric pressure) at which the boiling (or distillation) of a hydrocarbon liquid commences, proceeds, and finishes.
BOTTOMS Tower bottoms are residue remaining in a distillation unit after the highest boiling-point material to be distilled has been removed. Tank bottoms are the heavy materials that accumulate in the bottom of storage tanks, usually comprised of oil, water, and foreign matter.
BUBBLE TOWER A fractionating (distillation) tower in which the rising vapors pass through layers of condensate, bubbling under caps on a series of plates.
CATALYST A material that aids or promotes a chemical reaction between other substances but does not react itself. Catalysts increase reaction speeds and can provide control by increasing desirable reactions and decreasing undesirable reactions.
CATALYTIC CRACKING The process of breaking up heavier hydrocarbon molecules into lighter hydrocarbon fractions by use of heat and catalysts.
CAUSTIC WASH A process in which distillate is treated with sodium hydroxide to remove acidic contaminants that contribute to poor odor and stability.
CHD UNIT See Hydrodesulfurization.
COKE A high carbon-content residue remaining from the destructive distillation of petroleum residue.
COKING A process for thermally converting and upgrading heavy residual into lighter products and by-product petroleum coke. Coking also is the removal of all lighter distillable hydrocarbons that leaves a residue of carbon in the bottom of units or as buildup or deposits on equipment and catalysts.
CONDENSATE The liquid hydrocarbon resulting from cooling vapors.
CONDENSER A heat-transfer device that cools and condenses vapor by removing heat via a cooler medium such as water or lower-temperature hydrocarbon streams.
CONDENSER REFLUX Condensate that is returned to the original unit to assist in giving increased conversion or recovery.
COOLER A heat exchanger in which hot liquid hydrocarbon is passed through pipes immersed in cool water to lower its temperature.
CRACKING The breaking up of heavy molecular weight hydrocarbons into lighter hydrocarbon molecules by the application of heat and pressure, with or without the use of catalysts.
CRUDE ASSAY A procedure for determining the general distillation and quality characteristics of crude oil.

CRUDE OIL A naturally occurring mixture of hydrocarbons that usually includes small quantities of sulfur, nitrogen, and oxygen derivatives of hydrocarbons as well as trace metals.
CYCLE GAS OIL Cracked gas oil returned to a cracking unit.
DEASPHALTING Process of removing asphaltic materials from reduced crude using liquid propane to dissolve nonasphaltic compounds.
DEBUTANIZER A fractionating column used to remove butane and lighter components from liquid streams.
DE-ETHANIZER A fractionating column designed to remove ethane and gases from heavier hydrocarbons.
DEHYDROGENATION A reaction in which hydrogen atoms are eliminated from a molecule. Dehydrogenation is used to convert ethane, propane, and butane into olefins (ethylene, propylene, and butenes).
DEPENTANIZER A fractionating column used to remove pentane and lighter fractions from hydrocarbon streams.
DEPROPANIZER A fractionating column for removing propane and lighter components from liquid streams.
DESALTING Removal of mineral salts (most chlorides, e.g., magnesium chloride and sodium chloride) from crude oil.
DESULFURIZATION A chemical treatment to remove sulfur or sulfur compounds from hydrocarbons.
DEWAXING The removal of wax from petroleum products (usually lubricating oils and distillate fuels) by solvent absorption, chilling, and filtering.
DIETHANOLAMINE A chemical (C4H11O2N) used to remove H2S from gas streams.
DISTILLATE The products of distillation formed by condensing vapors.
DOWNFLOW Process in which the hydrocarbon stream flows from top to bottom.
DRY GAS Natural gas with so little natural gas liquids that it is nearly all methane with some ethane.
FEEDSTOCK Stock from which material is taken to be fed (charged) into a processing unit.
FLASHING The process in which a heated oil under pressure is suddenly vaporized in a tower by reducing pressure.
FLASH POINT Lowest temperature at which a petroleum product will give off sufficient vapor so that the vapor-air mixture above the surface of the liquid will propagate a flame away from the source of ignition.
FLUX Lighter petroleum used to fluidize heavier residual so that it can be pumped.
FOULING Accumulation of deposits in condensers, exchangers, etc.

FRACTION One of the portions of fractional distillation having a restricted boiling range.
FRACTIONATING COLUMN Process unit that separates various fractions of petroleum by simple distillation, with the column tapped at various levels to separate and remove fractions according to their boiling ranges.
FUEL GAS Refinery gas used for heating.
GAS OIL Middle-distillate petroleum fraction with a boiling range of about 350°-750° F, usually includes diesel fuel, kerosene, heating oil, and light fuel oil.
GASOLINE A blend of naphthas and other refinery products with sufficiently high octane and other desirable characteristics to be suitable for use as fuel in internal combustion engines.
HEADER A manifold that distributes fluid from a series of smaller pipes or conduits.
HEAT As used in the Health Considerations paragraphs of this document, heat refers to thermal burns for contact with hot surfaces, hot liquids and vapors, steam, etc.
HEAT EXCHANGER Equipment to transfer heat between two flowing streams of different temperatures. Heat is transferred between liquids or liquids and gases through a tubular wall.
HIGH-LINE OR HIGH-PRESSURE GAS High-pressure (100 psi) gas from cracking unit distillate drums that is compressed and combined with low-line gas as gas absorption feedstock.
HYDROCRACKING A process used to convert heavier feedstock into lower-boiling, higher-value products. The process employs high pressure, high temperature, a catalyst, and hydrogen.
HYDRODESULFURIZATION A catalytic process in which the principal purpose is to remove sulfur from petroleum fractions in the presence of hydrogen.
HYDROFINISHING A catalytic treating process carried out in the presence of hydrogen to improve the properties of low viscosity-index naphthenic and medium viscosity-index naphthenic oils. It is also applied to paraffin waxes and microcrystalline waxes for the removal of undesirable components. This process consumes hydrogen and is used in lieu of acid treating.
HYDROFORMING Catalytic reforming of naphtha at elevated temperatures and moderate pressures in the presence of hydrogen to form high-octane BTX aromatics for motor fuel or chemical manufacture. This process results in a net production of hydrogen and has rendered thermal reforming somewhat obsolete. It represents the total effect of numerous simultaneous reactions such as cracking, polymerization, dehydrogenation, and isomerization.
HYDROGENATION The chemical addition of hydrogen to a material in the presence of a catalyst.

INHIBITOR Additive used to prevent or retard undesirable changes in the quality of the product, or in the condition of the equipment in which the product is used.
ISOMERIZATION A reaction that catalytically converts straight-chain hydrocarbon molecules into branched-chain molecules of substantially higher octane number. The reaction rearranges the carbon skeleton of a molecule without adding or removing anything from the original material.
ISO-OCTANE A hydrocarbon molecule (2,2,4-trimethylpentane) with excellent antiknock characteristics on which the octane number of 100 is based.
KNOCKOUT DRUM A vessel wherein suspended liquid is separated from gas or vapor.
LEAN OIL Absorbent oil fed to absorption towers in which gas is to be stripped. After absorbing the heavy ends from the gas, it becomes fat oil. When the heavy ends are subsequently stripped, the solvent again becomes lean oil.
LOW-LINE or LOW-PRESSURE GAS Low-pressure (5 psi) gas from atmospheric and vacuum distillation recovery systems that is collected in the gas plant for compression to higher pressures.
NAPHTHA A general term used for low boiling hydrocarbon fractions that are a major component of gasoline. Aliphatic naphtha refers to those naphthas containing less than 0.1% benzene and with carbon numbers from C3 through C16. Aromatic naphthas have carbon numbers from C6 through C16 and contain significant quantities of aromatic hydrocarbons such as benzene (>0.1%), toluene, and xylene.
NAPHTHENES Hydrocarbons (cycloalkanes) with the general formula CnH2n, in which the carbon atoms are arranged to form a ring.
OCTANE NUMBER A number indicating the relative antiknock characteristics of gasoline.
OLEFINS A family of unsaturated hydrocarbons with one carbon-carbon double bond and the general formula CnH2n.
PARAFFINS A family of saturated aliphatic hydrocarbons (alkanes) with the general formula CnH2n+2.
POLYFORMING The thermal conversion of naphtha and gas oils into high-quality gasoline at high temperatures and pressure in the presence of recirculated hydrocarbon gases.
POLYMERIZATION The process of combining two or more unsaturated organic molecules to form a single (heavier) molecule with the same elements in the same proportions as in the original molecule.
PREHEATER Exchanger used to heat hydrocarbons before they are fed to a unit.

PRESSURE-REGULATING VALVE A valve that releases or holds process-system pressure (that is, opens or closes) either by preset spring tension or by actuation by a valve controller to assume any desired position between fully open and fully closed.
PYROLYSIS GASOLINE A by-product from the manufacture of ethylene by steam cracking of hydrocarbon fractions such as naphtha or gas oil.
PYROPHORIC IRON SULFIDE A substance typically formed inside tanks and processing units by the corrosive interaction of sulfur compounds in the hydrocarbons and the iron and steel in the equipment. On exposure to air (oxygen) it ignites spontaneously.
QUENCH OIL Oil injected into a product leaving a cracking or reforming heater to lower the temperature and stop the cracking process.
RAFFINATE The product resulting from a solvent extraction process and consisting mainly of those components that are least soluble in the solvents. The product recovered from an extraction process is relatively free of aromatics, naphthenes, and other constituents that adversely affect physical parameters.
REACTOR The vessel in which chemical reactions take place during a chemical conversion type of process.
REBOILER An auxiliary unit of a fractionating tower designed to supply additional heat to the lower portion of the tower.
RECYCLE GAS High hydrogen-content gas returned to a unit for reprocessing.
REDUCED CRUDE A residual product remaining after the removal by distillation of an appreciable quantity of the more volatile components of crude oil.
REFLUX The portion of the distillate returned to the fractionating column to assist in attaining better separation into desired fractions.
REFORMATE An upgraded naphtha resulting from catalytic or thermal reforming.
REFORMING The thermal or catalytic conversion of petroleum naphtha into more volatile products of higher octane number. It represents the total effect of numerous simultaneous reactions such as cracking, polymerization, dehydrogenation, and isomerization.
REGENERATION In a catalytic process the reactivation of the catalyst, sometimes done by burning off the coke deposits under carefully controlled conditions of temperature and oxygen content of the regeneration gas stream.
SCRUBBING Purification of a gas or liquid by washing it in a tower.
SOLVENT EXTRACTION The separation of materials of different chemical types and solubilities by selective solvent action.

SOUR GAS Natural gas that contains corrosive, sulfur-bearing compounds such as hydrogen sulfide and mercaptans.
STABILIZATION A process for separating the gaseous and more volatile liquid hydrocarbons from crude petroleum or gasoline and leaving a stable (less-volatile) liquid so that it can be handled or stored with less change in composition.
STRAIGHT-RUN GASOLINE Gasoline produced by the primary distillation of crude oil. It contains no cracked, polymerized, alkylated, reformed, or visbroken stock.
STRIPPING The removal (by steam-induced vaporization or flash evaporation) of the more volatile components from a cut or fraction.
SULFURIC ACID TREATING A refining process in which unfinished petroleum products such as gasoline, kerosene, and lubricating oil stocks are treated with sulfuric acid to improve their color, odor, and other characteristics.
SULFURIZATION Combining sulfur compounds with petroleum lubricants.
SWEETENING Processes that either remove obnoxious sulfur compounds (primarily hydrogen sulfide, mercaptans, and thiophens) from petroleum fractions or streams, or convert them, as in the case of mercaptans, to odorless disulfides to improve odor, color, and oxidation stability.
SWITCH LOADING The loading of a high static-charge retaining hydrocarbon (i.e., diesel fuel) into a tank truck, tank car, or other vessel that has previously contained a low-flash hydrocarbon (gasoline) and may contain a flammable mixture of vapor and air.
TAIL GAS The lightest hydrocarbon gas released from a refining process.
THERMAL CRACKING The breaking up of heavy oil molecules into lighter fractions by the use of high temperature without the aid of catalysts.
TURNAROUND A planned complete shutdown of an entire process or section of a refinery, or of an entire refinery to perform major maintenance, overhaul, and repair operations and to inspect, test, and replace process materials and equipment.
VACUUM DISTILLATION The distillation of petroleum under vacuum which reduces the boiling temperature sufficiently to prevent cracking or decomposition of the feedstock.
VAPOR The gaseous phase of a substance that is a liquid at normal temperature and pressure.
VISBREAKING Viscosity breaking is a low-temperature cracking process used to reduce the viscosity or pour point of straight-run residuum.
WET GAS A gas containing a relatively high proportion of hydrocarbons that are recoverable as liquids.

NATURAL GAS GLOSSARY

ANNUALLY - means at intervals not exceeding 15 months, but at least once each calendar year.

Balancing item: Represents differences between the sum of the components of natural gas supply and the sum of the components of natural gas disposition. These differences may be due to quantities lost or to the effects of data reporting problems. Reporting problems include differences due to the net result of conversions off low data metered at varying temperature and pressure bases and converted to a standard temperature and pressure base; the effect of variations in company accounting and billing practices; differences between billing cycle and calendar period time frames; and imbalances resulting from the merger of data reporting systems that vary in scope, format, definitions, and type of respondents.

Base gas: The quantity of natural gas needed to maintain adequate reservoir pressures and deliverability rates throughout the withdrawal season. Base gas usually is not withdrawn and remains in the reservoir. All natural gas native to a depleted reservoir is included in the base gas volume.

Biomass: Organic nonfossil material of biological origin constituting a renewable energy source.

British thermal unit: The quantity of heat required to raise the temperature of 1 pound of liquid water by 1 degree Fahrenheit at the temperature at which water has its greatest density (approximately 39 degrees Fahrenheit).

CATHODIC (CORROSION) PROTECTION - a procedure by which underground metallic pipe is protected against deterioration (rusting and pitting).

Citygate: A point or measuring station at which a distributing gas utility receives gas from a natural gas pipeline company or transmission system.

Coke oven gas: The mixture of permanent gases produced by the carbonization of coal in a coke oven at temperatures in excess of 1,000 degrees Celsius.

Compressed natural gas (CNG): **Natural gas** compressed to a pressure at or above 200-248 bar (i.e., 2900-3600 pounds per square inch) and stored in high-pressure containers. It is used as a fuel for natural gas-powered vehicles.

Condensate (lease condensate): Light liquid hydrocarbons recovered from lease separators or field facilities at associated and non-associated natural gas wells. Mostly pentanes and heavier hydrocarbons. Normally enters the crude oil stream after production.

CUSTOMER METER – a device that measures the volume of gas transferred from an operator to the consumer.

Delivered (gas): The physical transfer of natural, synthetic, and/or supplemental gas from facilities operated by the responding company to facilities operated by others or to consumers.

Depleted storage field: A sub-surface natural geological reservoir, usually a depleted gas or oil field, used for storing natural gas.

DOWNSTREAM – any point in the direction of flow of a gas from the reference point.

Dry natural gas: Natural gas which remains after: 1) the liquefiable hydrocarbon portion has been removed from the gas stream (i.e., gas after lease, field, and/or plant separation); and 2) any volumes of nonhydrocarbon gases have been removed where they occur in sufficient quantity to render the gas unmarketable. Note: Dry natural gas is also known as consumer-grade natural gas. The parameters for measurement are cubic feet at 60 degrees Fahrenheit and 14.73 pounds per square inch absolute. Also see **Natural gas.**

Dry natural gas production: The process of producing consumer-grade natural gas. Natural gas withdrawn from reservoirs is reduced by volumes used at the production (lease) site and by processing losses. Volumes used at the production site include (1) the volume returned to reservoirs in cycling, repressuring of oil reservoirs, and conservation operations; and (2) gas vented and flared. Processing losses include (1) nonhydrocarbon gases (e.g., water vapor, carbon dioxide, helium, hydrogen sulfide, and nitrogen) removed from the gas stream; and (2) gas converted to liquid form, such as lease condensate and plant liquids. Volumes of dry gas withdrawn from gas storage reservoirs are not considered part of production. Dry natural gas production equals marketed production less extraction loss.

Electric power sector: An energy-consuming sector that consists of electricity only and combined heat and power(CHP) plants whose primary business is to sell electricity, or electricity and heat, to the public.

Electric utility: A corporation, person, agency, authority, or other legal entity or instrumentality aligned with distribution facilities for delivery of electric energy for use primarily by the public. Included are investor-owned electric utilities, municipal and State utilities, Federal electric utilities, and rural electric cooperatives. A few entities that are tariff based and corporately aligned with companies that own distribution facilities are also included.

EMERGENCY PLAN – written procedures for responding to emergencies on the pipeline system.

Exports: Shipments of goods from within the 50 States and the District of Columbia to U.S. possessions and territories or to foreign countries.

Extraction loss: See **Natural gas plant liquids (NGPL) production**.

Flare: A tall stack equipped with burners used as a safety device at wellheads, refining facilities, gas processing plants, and chemical plants. Flares are used for the combustion and disposal of combustible gases. The gases are piped to a remote, usually elevated, location and burned in an open flame in the open air using a specially designed burner tip, auxiliary fuel, and steam or air. Combustible gases are flared most often due to emergency relief, overpressure, process upsets, startups, shutdowns, and other operational safety reasons. Natural gas that is uneconomical for sale is also flared. Often natural gas is flared as a result of the unavailability of a method for transporting such gas to markets.

Gas Condensate Well Gas: Natural gas remaining after the removal of the lease condensate.

GAS OPERATOR – a gas operator may be a gas utility company, a municipality, or an individual operating a housing project, apartment complex, condominium, or a mobile home park served by a master meter. The operator is ultimately responsible for complying with the pipeline safety regulations.

Gas well: A well completed for production of natural gas from one or more gas zones or reservoirs. Such wells contain no completions for the production of crude oil.

Gross withdrawals: Full well stream volume from both oil and gas wells, including all natural gas plant liquids and nonhydrocarbon gases after oil, lease condensate, and water have been removed. Also includes production delivered as royalty payments and production used as fuel on the lease.

Heating value: The average number of British thermal units per cubic foot of natural gas as determined from tests of fuel samples.

HIGH-PRESSURE DISTRIBUTION SYSTEM – a distribution system in which the gas pressure in the main is higher than the pressure provided to the customer; therefore, a pressure regulator is required on each service to control pressure to the customer.

Hydrocarbon gas liquids (HGL): A group of hydrocarbons including ethane, propane, normal butane, isobutane, and natural gasoline, and their associated olefins, including ethylene, propylene, butylene, and isobutylene. As marketed products, HGL represents all natural gas liquids (NGL) and olefins. EIA reports production of HGL from refineries (liquefied refinery gas, or LRG) and natural gas plants (natural gas plant liquids, or NGPL). Excludes liquefied natural gas (LNG).

Imports: Receipts of goods into the 50 States and the District of Columbia from U.S. possessions and territories or from foreign countries.

INCIDENT – an event that involves a release of natural gas from a pipeline facility that results in: (1) a death or personal injury necessitating in-patient hospitalization; (2) estimated property damage of $50,000 or more; or (3) an event that the operator deems significant.

Intransit deliveries: Redeliveries to a foreign country of foreign gas received for transportation across U.S. territory, and deliveries of U.S. gas to a foreign country for transportation across its territory and redelivery to the United States.

Intransit receipts: Receipts of foreign gas for transportation across U.S. territory and redelivery to a foreign country, and redeliveries to the United States of U.S. gas transported across foreign territory.

Lease and plant fuel: Natural gas used in well, field, and lease operations (such as gas used in drilling operations, heaters, dehydrators, and field compressors) and as fuel in natural gas processing plants.

Lease fuel: Natural gas used in well, field, and lease operations, such as gas used in drilling operations, heaters, dehydrators, and field compressors.

Lease separator: A facility installed at the surface for the purpose of separating the full well stream volume into two or three parts at the temperature and pressure conditions set by the separator. For oil wells, these parts include produced crude oil, natural gas, and water. For gas wells, these parts include produced natural gas, lease condensate, and water.

Liquefied natural gas (LNG): Natural gas (primarily methane) that has been liquefied by reducing its temperature to -260 degrees Fahrenheit at atmospheric pressure.

LOW-PRESSURE DISTRIBUTION SYSTEM – a distribution system in which the gas pressure in the main is substantially the same as the pressure provided to the customer; normally a pressure regulator is not required on individual service lines.

MAIN – a natural gas distribution pipeline that serves as a common source of supply for more than one service line.

Manufactured gas: A gas obtained by destructive distillation of coal or by the thermal decomposition of oil, or by the reaction of steam passing through a bed of heated coal or coke. Examples are coal gases, coke oven gases, producer gas, blast furnace gas, blue (water) gas, carbureted water gas. Btu content varies widely.

Marketed production: Gross withdrawals less gas used for repressuring, quantities vented and flared, and nonhydrocarbon gases removed in treating or processing operations. Includes all quantities of gas used in field and processing plant operations.
MASTER METER SYSTEM – a natural gas pipeline system for distributing natural gas for resale within, but not limited to, a distinct area, such as a mobile home park, housing project, or apartment complex, where the operator purchases metered gas from an outside source. The natural gas distribution pipeline system supplies the ultimate consumer who either purchases the gas directly through a meter or by other means such as by rent.
MAXIMUM ALLOWABLE OPERATING PRESSURE (MAOP) – the maximum pressure at which a pipeline may be operated in compliance with the gas pipeline safety regulations. It is established by design, past operating history, pressure testing, and pressure ratings of components.
MUNICIPALITY – a city, county, or any other political subdivision of a state.
Native gas: Gas in place at the time that a reservoir was converted to use as an underground storage reservoir in contrast to injected gas volumes.
Natural gas: a non-toxic, colorless fuel, about one-third lighter than air. Natural gas burns only when mixed with air in certain proportions and ignited by a source of ignition (spark or flame). Natural gas in its natural state may not have an odor.
Natural gas field facility: A field facility designed to process natural gas produced from more than one lease for the purpose of recovering condensate from a stream of natural gas; however, some field facilities are designed to recover propane, normal butane, pentanes plus, etc., and to control the quality of natural gas to be marketed.
Natural gas gross withdrawals: Full well-stream volume of produced natural gas, excluding condensate separated at the lease.
Natural gas hydrates: Solid, crystalline, wax-like substances composed of water, methane, and usually a small amount of other gases, with the gases being trapped in the interstices of a water-ice lattice. They form beneath permafrost and on the ocean floor under conditions of moderately high pressure and at temperatures near the freezing point of water.
Natural gas lease production: Gross withdrawals of natural gas minus gas production injected on the lease into producing reservoirs, vented, flared, used as fuel on the lease, and nonhydrocarbon gases removed in treating or processing operations on the lease.
Natural Gas Liquids (NGL): A group of hydrocarbons including ethane, propane, normal butane, isobutane, and

natural gasoline. Generally include natural gas plant liquids and all liquefied refinery gases except olefins.

Natural gas liquids production: The volume of natural gas liquids removed from natural gas in lease separators, field facilities, gas processing plants, or cycling plants during the report year.

Natural gas marketed production: Gross withdrawals of natural gas from production reservoirs, less gas used for reservoir repressuring, nonhydrocarbon gases removed in treating and processing operations, and quantities vented and flared.

Natural gas marketer: A company that arranges purchases and sales of natural gas. Unlike pipeline companies or local distribution companies, a marketer does not own physical assets commonly used in the supply of natural gas, such as pipelines or storage fields. A marketer may be an affiliate of another company, such as a local distribution company, natural gas pipeline, or producer, but it operates independently of other segments of the company. In States with residential choice programs, marketers serve as alternative suppliers to residential users of natural gas, which is delivered by a local distribution company.

Natural gas plant liquids (NGPL): Those hydrocarbons in natural gas that are separated as liquids at natural gas processing, fractionating, and cycling plants. Products obtained include ethane, liquefied petroleum gases (propane, normal butane, and isobutane), and natural gasoline. Component products may be fractionated or mixed. Lease condensate and **plant condensate** are excluded. Note: Some EIA publications categorize NGPL production as field production, in accordance with definitions used prior to January 2014.

Natural gas plant liquids (NGPL) production: The extraction of gas plant liquids constituents such as ethane, propane, normal butane, isobutane, and natural gasoline, sometimes referred to as extraction loss. Usually reported in barrels or gallons, but may be reported in cubic feet for purposes of comparison with dry natural gas volumes.

Natural Gas Policy Act of 1978 (NGPA): Signed into law on November 9, 1978, the NGPA is a framework for the regulation of most facets of the natural gas industry.

Natural gas processing plant: Facilities designed to recover natural gas liquids from a stream of natural gas that may or may not have passed through lease separators and/or field separation facilities. These facilities control the quality of the natural gas to be marketed. Cycling plants are classified as gas processing plants.

Natural gas production: See **Dry natural gas production**.

Natural Gas Used for Injection: Natural gas used to pressurize crude oil reservoirs in an attempt to increase oil recovery or in instances where there is no market for the natural gas. Natural gas used for injection is sometimes referred to as repressuring.

Natural gas utility demand-side management (DSM) program sponsor: A DSM (demand-side management) program sponsored by a natural gas utility that suggests ways to increase the energy efficiency of buildings, to reduce energy costs, to change the usage patterns, or to promote the use of a different energy source.

Natural gas, "dry": See **Dry natural gas**.

Natural gasoline: A commodity product commonly traded in NGL markets that comprises liquid hydrocarbons (mostly pentanes and hexanes) and generally remains liquid at ambient temperatures and atmospheric pressure. Natural gasoline is equivalent to **pentanes plus**.

Natural Gasoline and Isopentane: A mixture of hydrocarbons, mostly pentanes and heavier, extracted from natural gas, that meets vapor pressure, end-point, and other specifications for natural gasoline set by the Gas Processors Association. Includes isopentane which is a saturated branch-chain hydrocarbon, (C_5H_{12}), obtained by fractionation of natural gasoline or isomerization of normal pentane.

Nonhydrocarbon gases: Typical nonhydrocarbon gases that may be present in reservoir natural gas, such as carbondioxide, helium, hydrogen sulfide, and nitrogen.

Nonutility power producer: A corporation, person, agency, authority, or other legal entity or instrumentality that owns or operates facilities for electric generation and is not an electric utility. Nonutility power producers include qualifying cogenerators, qualifying small power producers, and other nonutility generators (including independent power producers). Non-utility power producers are without a designated franchised service area and do not file forms listed in the Code of Federal Regulations, Title 18, Part 141

Offshore reserves and production: Unless otherwise dedicated, reserves and production that are in either state or Federal domains, located seaward of the coastline.

Oil well: A well completed for the production of crude oil from at least one oil zone or reservoir.

Olefinic hydrocarbons (olefins): Unsaturated hydrocarbon compounds with the general formula CnH2n containing at least one carbon-to-carbon double-bond. Olefins are produced at crude oil refineries and petrochemical plants and are not naturally occurring constituents of oil and natural gas. Sometimes referred to as alkenes or unsaturated hydrocarbons. Excludes aromatics.

On-system sales: Sales to customers where the delivery point is a point on, or directly interconnected with, a transportation, storage, and/or distribution system operated by the reporting company.

OPERATING AND MAINTENANCE PLAN - written procedures for operations and maintenance on natural gas pipeline systems.

Outer Continental Shelf: Offshore Federal domain.

OVERPRESSURE PROTECTION EQUIPMENT - equipment installed to protect and prevent pressure in a system from exceeding the maximum allowable operating pressure (MAOP).

Paraffinic hydrocarbons: Saturated hydrocarbon compounds with the general formula C_nH_{2n+2} containing only single bonds. Sometimes referred to as alkanes or natural gas liquids.

Pipeline (natural gas): A continuous pipe conduit, complete with such equipment as valves, compressor stations, communications systems, and meters for transporting natural and/or supplemental gas from one point to another, usually from a point in or beyond the producing field or processing plant to another pipeline or to points of utilization. Also refers to a company operating such facilities.

Pipeline fuel: Gas consumed in the operation of pipelines, primarily in compressors.

PRESSURE REGULATING/RELIEF STATION - a device to automatically reduce and control the gas pressure in a pipeline downstream from a higher pressure source of natural gas. It includes any enclosures, relief devices, ventilating equipment, and any piping and auxiliary equipment, such as valves, regulators, control instruments, or control lines.

PRETESTED PIPE - pipe that has been tested by the operator to 100 psig for at least one hour.

Production, natural gas, wet after lease separation: The volume of natural gas withdrawn from reservoirs less (1) the volume returned to such reservoirs in cycling, repressuring of oil reservoirs, and conservation operations; less (2) shrinkage resulting from the removal of lease condensate; and less (3) nonhydrocarbon gases where they occur in sufficient quantity to render the gas unmarketable. Note: Volumes of gas withdrawn from gas storage reservoirs and native gas that has been transferred to the storage category are not considered part of production. This production concept is not the same as marketed production, which excludes vented and flared gas.

Propane air: A mixture of propane and air resulting in a gaseous fuel suitable for pipeline distribution.

Proved energy reserves: Estimated quantities of energy sources that analysis of geologic and engineering data demonstrates with reasonable certainty are recoverable under existing economic and operating conditions. The location, quantity, and grade of the energy source are usually considered to be well established in such reserves. Note: This term is equivalent to "Measured Reserves" as defined in the resource/reserve classification contained in the U.S. Geological Survey Circular 831, 1980. Measured and indicated reserves, when combined, constitute demonstrated reserves.

Receipts:

- Deliveries of fuel to an electric plant
- Purchases of fuel
- All revenues received by an exporter for the reported quantity exported

Refinery gas: **Still gas** consumed as refinery fuel.

Refinery olefins: Subset of olefinic hydrocarbons (olefins) produced at crude oil refineries, including ethylene, propylene, butylene, and isobutylene.

Repressuring: The injection of gas into oil or gas formations to effect greater ultimate recovery.

SERVICE LINE - a natural gas distribution line that transports gas from a common source of supply to a customer's meter, or to the connection to a customer's piping if the piping is farther downstream or if there is no meter.

SERVICE REGULATOR - a device designed to reduce and limit the gas pressure provided to a customer.

SERVICE RISER - the section of a service line which extends out of the ground and is often near the wall of a building. This usually includes a shut-off valve and a service regulator.

Shale Gas: Natural gas produced from wells that are open to shale formations. Shale is a fine-grained, sedimentary rock composed of mud from flakes of clay minerals and tiny fragments (silt-sized particles) of other materials. The shale acts as both the source and the reservoir for the natural gas.

SHUT-OFF VALVE - a valve used to stop the flow of gas. The valve may be located upstream of the service regulator or below ground at the property line or where the service line connects to the main.

Supplemental gaseous fuels supplies: Synthetic natural gas, propane-air, coke oven gas, refinery gas, biomass gas, air injected for Btu stabilization, and manufactured gas commingled and distributed with natural gas.

Synthetic natural gas (SNG): (Also referred to as substitute natural gas) A manufactured product, chemically similar in most respects to natural gas, resulting from the conversion or reforming of hydrocarbons that may easily be substituted for or interchanged with pipeline-quality natural gas.

Therm: One hundred thousand (100,000) Btu.

Total Natural Gas Storage Field Capacity (Design Capacity): The maximum quantity of natural gas (including both base gas and working gas) that can be stored in a natural gas underground storage facility in accordance with its design specifications, the physical characteristics of the reservoir, installed compression equipment, and operating procedures particular to the site. Reported storage field capacity data are reported in thousand cubic feet at standard temperature and pressure.

Unaccounted for (natural gas): Represents differences between the sum of the components of natural gas supply and the sum of components of natural gas disposition. These differences maybe due to quantities lost or to the effects of data reporting problems. Reporting problems include differences due to the net result of conversions of flow data metered at varying temperatures and pressure bases and converted to a standard temperature and pressure base; the effect of variations in company accounting and billing practices; differences between billing cycle and calendar-period time frames; and imbalances resulting from the merger of data reporting systems that vary in scope, format, definitions, and type of respondents.

Underground natural gas storage: The use of sub-surface facilities for storing natural gas for use at a later time. The facilities are usually hollowed-out salt domes, geological reservoirs (depleted oil or gas fields) or water-bearing sands (called aquifers) topped by an impermeable cap rock.

Underground natural gas storage injections: Natural gas put (injected) into underground storage reservoirs.

Underground storage withdrawals: Natural gas removed from underground storage reservoirs.

Unit value, consumption: Total price per specified unit, including all taxes, at the point of consumption.

Unit value, wellhead: The wellhead sales price, including charges for natural gas plant liquids subsequently removed from the gas; gathering and compression charges; and state production, severance, and/or similar charges.

UPSTREAM - from a reference point, any point located nearest the origin of flow, that is, before the reference point is reached.

Vehicle fuel consumption: Vehicle fuel consumption is computed as the vehicle miles traveled divided by the fuel efficiency reported in miles per gallon (MPG). Vehicle fuel consumption is derived from the actual vehicle mileage collected and the assigned MPGs obtained from EPA certification files adjusted for on-road driving. The quantity of fuel used by vehicles.

Vented: Natural gas that is disposed of by releasing to the atmosphere.

Vented natural gas: See **vented**.

Wellhead: The point at which the crude (and/or natural gas) exits the ground. Following historical precedent, the volume and price for crude oil production are labeled as "wellhead, "even though the cost and volume are now generally measured at the lease boundary. In the context of domestic crude price data, the term "wellhead" is the generic term used to reference the production site or lease property.
Wellhead price: The value at the mouth of the well. In general, the wellhead price is considered to be the sales price obtainable from a third party in an arm's length transaction. Posted prices, requested prices, or prices as defined by lease agreements, contracts, or tax regulations should be used where applicable.
Working gas: The quantity of natural gas in the reservoir that is in addition to the cushion or base gas. It may or may not be completely withdrawn during any particular withdrawal season. Conditions permitting, the total working capacity could be used more than once during any season. Volumes of working gas are reported in thousand cubic feet at standard temperature and pressure.

COMMONLY ABBREVIATED ORGANIZATION/ACRONYMS

AGA - American Gas Association.
ANSI - American National Standards Institute, formerly the United States of America Standards Institute (USASI). All current standards issued by USASI and American Standard Association (ASA) have been redesignated as American National Standards Institute and continue in effect.
APGA - American Public Gas Association.
API - American Petroleum Institute.
ASME - American Society of Mechanical Engineers.
ASTM - American Society for Testing and Materials.
DOT - U.S. Department of Transportation.
GPTC - Gas Piping Technology Committee.
INGAA - Interstate Natural Gas Association of America.
MEA - Midwest Energy Association.
MSS - Manufacturers Standardization Society of the Valve and Fittings Industry.
NACE - National Association of Corrosion Engineers. (NACE International)
NFPA - National Fire Protection Association.
OPS - Office of Pipeline Safety. The pipeline safety division of the DOT's Research and Special Programs Administration. For addresses of OPS regional offices, see the attached list of agencies and organizations.

RSPA – Research and Special Programs Administration. A major subdivision of the DOT, it includes the Office of Pipeline Safety. For addresses of regional offices, see the enclosed handout.

SGA – Southern Gas Association.

! *Disclaimer: this information is advisory in nature and is not intended to identify all scenarios or situations a person might encounter.*

! *Following these guidelines will not guarantee your safety.*

Floods

When indoors during a flood

- Listen to the radio or television for information.
- Be aware that flash flooding can occur anywhere.
- If there is any possibility of a flash flood where you are, move immediately to higher ground. Do not wait for instructions to move.
- Be aware of nearby streams, drainage channels, canyons, and other areas known to flood suddenly. Flash floods can occur in these areas with or without such typical warnings as rain clouds or heavy rain.

When evacuating

- Do not touch electrical equipment if you are wet or standing in water.
- Do not walk through moving water. Six inches of moving water can make you fall. If you have to walk in water, walk where the water is not moving. Use a stick to check the firmness of the ground in front of you.

Driving in flood conditions

- Do not drive into flooded areas.
- If floodwaters rise around your car, abandon the car and move to higher ground if you can do so safely. You and the vehicle can be quickly swept away. Nearly half of all flash flood deaths are vehicle-related.
- Avoid driving through even low levels of water. Six inches of water will reach the bottom of most passenger cars causing loss of control and possible stalling. A foot of water will float many vehicles. Two feet of rushing water can carry away most vehicles including sport utility vehicles (SUV's) and pick-ups.
- Be especially careful at night when flash floods are harder to recognize.
- If your vehicle becomes caught in high water and stalls, leave it immediately and seek higher ground if you can do so safely. Rapidly rising water can sweep a vehicle and its occupants away.

After a flood

- Avoid floodwaters; water may be contaminated by oil, gasoline, or raw sewage and may be electrically charged from underground or downed power lines.

- Avoid moving water. Be aware of areas where floodwaters have receded. Roads may have weakened and could collapse under the weight of a car.
- Stay away from downed power lines, and report them to the power company.
- Stay out of any building if it is surrounded by floodwaters. Use extreme caution when entering buildings; there may be hidden damage, particularly in foundations.
- Service damaged septic tanks, cesspools, pits, and leaching systems as soon as possible. Damaged sewage systems are serious health hazards.

Flood conditions--terms

- Flood Watch: Flooding is possible. Tune in to NOAA Weather Radio, commercial radio, or television for information.
- Flash Flood Watch: Flash flooding is possible. Be prepared to move to higher ground; listen to NOAA Weather Radio, commercial radio, or television for information.
- Flood Warning: Flooding is occurring or will occur soon; if advised to evacuate, do so immediately.
- Flash Flood Warning: A flash flood is occurring; seek higher ground on foot immediately.

Hurricanes

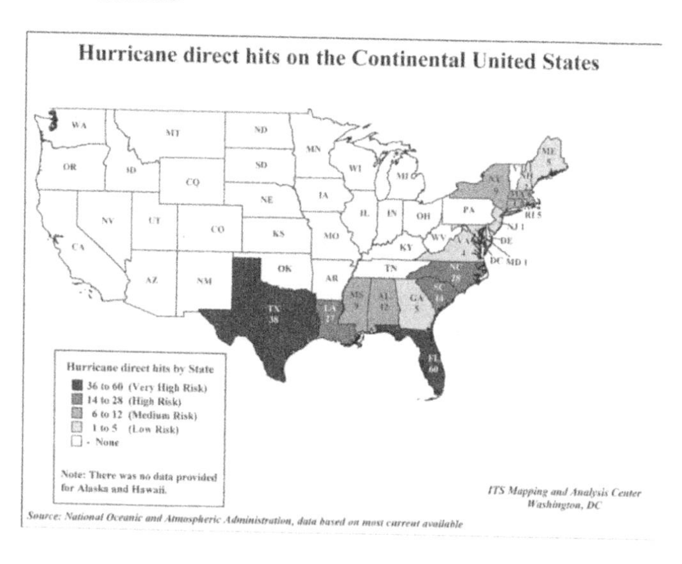

Evacuating

- You should evacuate under the following conditions:
- If you are directed by local authorities to do so. Be sure to follow their instructions.

- If you are in a temporary structure—such shelters are particularly hazardous during hurricanes no matter how well fastened to the ground.
- If you live on the coast, on a floodplain, near a river, or on an inland waterway.
- If you are unable to evacuate, go to your wind-safe room. If you do not have one, follow the guidelines below.

During a hurricane

- Stay indoors during the hurricane and away from windows and glass doors.
- Close all interior doors—secure and brace external doors.
- Keep curtains and blinds closed. Do not be fooled if there is a lull; it could be the eye of the storm—winds will pick up again.
- Take refuge in a small interior room, closet, or hallway on the lowest level and lie on the floor under a table or another sturdy object.

Hurricane terminology

- Tropical Depression: An organized system of clouds and thunderstorms with a defined surface circulation and maximum sustained winds of 38 MPH (33 knots) or less. Sustained winds are defined as one-minute average wind measured at about 33 ft. (10 meters) above the surface.
- Tropical Storm: An organized system of strong thunderstorms with a defined surface circulation and maximum sustained winds of 39-73 MPH (34-63 knots).
- Hurricane: An intense tropical weather system of strong thunderstorms with a well-defined surface circulation and maximum sustained winds of 74 MPH (64 knots) or higher.
- Storm Surge: A dome of water pushed onshore by hurricane and tropical storm winds. Storm surges can reach 25 feet high and be 50-100 miles wide.
- Storm Tide: A combination of storm surge and the normal tide (i.e., a 15-foot storm surge combined with a 2-foot normal high tide over the mean sea level creates a 17-foot storm tide). Hurricane/Tropical Storm Watch Hurricane/tropical storm conditions are possible in the specified area, usually within 36 hours. Tune in to NOAA Weather Radio, commercial radio, or television for information.
- Hurricane/Tropical Storm Warning: Hurricane/tropical storm conditions are expected in the specified area, usually within 24 hours. Short Term Watches and Warnings These warnings provide detailed information about specific hurricane threats, such as flash floods and tornadoes.
-

Saffir-Simpson Hurricane Scale

Catego ry	Category Description	Level of Damage
1	Wind Speed: 74 - 95 MPH Storm Surge: 4 - 5 Feet Above Normal	Primary damaged to unanchored mobile homes, shrubbery, and trees. Some coastal road flooding and minor pier damage. Little damage to building structures.
2	Wind Speed: 96 - 110 MPH Storm Surge: 6 - 8 Feet Above Normal	Considerable damage to mobile homes, piers, and vegetation. Coastal and low-lying escape routes flood 2 - 4 hours before arrival of hurricane center. Buildings sustain roofing material, door, and window damage. Small craft in unprotected moorings break moorings.
3	Wind Speed: 111 - 130 MPH Storm Surge: 9 - 12 Feet Above Normal	Mobile homes destroyed. Some structural damage to small homes and utility buildings. Flooding near coast destroys smaller structures; larger structures damaged by floating debris. Terrain continuously lower than 5 feet. ASL may be flooded up to 6 miles inland.
4	Wind Speed: 131 - 155 MPH Storm Surge: 13 - 18 Feet Above Normal	Extensive curtain wall failures with some complete roof structure failure on small residences. Major erosion of beaches. Major damage to lower floors of structures near the shore. Terrain continuously lower than 10 feet. ASL may flood (and require mass evacuations) up to 6 miles inland.
5	Wind Speed: Over 155 MPH Storm Surge: Over 18 Feet Above Normal	Complete roof failure on many homes and industrial buildings. Some complete building failures. Major damage to lower floors of all structures located less than 15 feet ASL and within 500 yards of the shoreline. Massive evacuation of low ground residential areas may be required.

Thunderstorms and Lightning

<table><tr><td></td><td>WARNING!
Remember the 30/30 lightning safety rule: Go indoors if, after seeing lightning, you cannot count to 30 before hearing thunder. Stay indoors for 30 minutes after hearing the last clap of thunder.</td></tr></table>

Before a thunderstorm

The following are guidelines for what you should do if a thunderstorm is likely in your area:

- Postpone outdoor activities.
- Get inside a building, or hard top automobile (not a convertible). Although you may be injured if lightning strikes your car, you are much safer inside a vehicle than outside.
- Remember, rubber-soled shoes and rubber tires provide NO protection from lightning. However, the steel frame of a hard-topped vehicle provides increased protection if you are not touching metal.
- Secure outdoor objects that could blow away or cause damage.
- Avoid showering or bathing. Plumbing and bathroom fixtures can conduct electricity.
- Use a corded telephone only for emergencies. Cordless and cellular telephones are safe to use.
- Unplug appliances and other electrical items such as computers and turn off air conditioners. Power surges from lightning can cause serious damage.
- Use your battery-operated NOAA Weather Radio for updates from local officials.

During a thunderstorm

Avoid the following:

- Natural lightning rods such as a tall, isolated tree in an open area
- Hilltops, open fields, the beach, or a boat on the water
- Isolated sheds or other small structures in open areas
- Anything metal—cranes, derricks, rigging, lifts etc.

Thunderstorms and lightning conditions—terms

- ***Severe thunderstorm watch:*** Tells you when and where severe thunderstorms are likely to occur. Watch the sky and stay tuned to NOAA Weather Radio, commercial radio, or television for information.

- ***Severe thunderstorm warning:*** Issued when severe weather has been reported by spotters or indicated by radar. Warnings indicate imminent danger to life and property to those in the path of the storm.

Tornadoes

A tornado is a violently rotating column of air extending from a thunderstorm to the ground. Tornadoes may appear nearly transparent until dust and debris are picked up or a cloud forms within the funnel. The average tornado moves from southwest to northeast, but tornadoes have been known to move in any direction. The average forward speed is 30 mph but may vary from nearly stationary to 70 mph. The strongest tornadoes have rotating winds of more than 250 mph. Tornadoes can accompany tropical storms and hurricanes as they move onto land. Waterspouts are tornadoes which form over warm water. They can move onshore and cause damage to coastal areas. Tornadoes can occur at any time of the year.

Tornadoes have occurred in every state, but they are most frequent east of the Rocky Mountains during the spring and summer months. In the southern states, peak tornado occurrence is March through May, while peak months in the northern states are during the late spring and summer. Tornadoes are most likely to occur between 3 and 9 p.m. but can happen at any time.

Before a tornado

- Be alert to changing weather conditions.
- Listen to NOAA Weather Radio or to commercial radio or television newscasts for the latest information.
- Look for approaching storms
- Look for the following danger signs:
- Dark, often greenish sky
- Large hail
- A large, dark, low-lying cloud (particularly if rotating)
- Loud roar, similar to a freight train.
- If you see approaching storms or any of the danger signs, be prepared to take shelter immediately.

During a tornado

If you are under a tornado warning, seek shelter immediately! Use the following guidelines to determine your safest option:

In a permanent building

- Go to a pre-designated shelter area such as a safe room, basement, storm cellar, or the lowest building level.

- If there is no basement, go to the center of an interior room on the lowest level (closet, interior hallway) away from corners, windows, doors, and outside walls. Put as many walls as possible between you and the outside.
- Get under a sturdy table and use your arms to protect your head and neck.
- Do not open windows.

In a vehicle, trailer, or temp building

- Get out immediately and go to the lowest floor of a sturdy, nearby building or a storm shelter. Mobile homes, even if tied down, offer little protection from tornadoes.
- Never try to outrun a tornado in urban or congested areas in a car or truck. Instead, leave the vehicle immediately for safe shelter.

Outside with no shelter

- Lie flat in a nearby ditch or depression and cover your head with your hands.
- Be aware of the potential for flooding.
- Do not get under an overpass or bridge. You are safer in a low, flat location.
- Do what you can to avoid flying debris. Flying debris from tornadoes causes most fatalities and injuries.

After a Tornado

Aiding the Injured

- Check for injuries. Do not attempt to move seriously injured persons unless they are in immediate danger of death or further injury. If you must move an unconscious person, first stabilize the neck and back, then call for help immediately.
- If the victim is not breathing, carefully position the victim for artificial respiration, clear the airway, and commence mouth-to-mouth resuscitation.
- Maintain body temperature with blankets. Be sure the victim does not become overheated.
- Never try to feed liquids to an unconscious person.

Health

- Be aware of exhaustion. Don't try to do too much at once. Set priorities and pace yourself. Get enough rest.
- Drink plenty of clean water. Eat well. Wear sturdy work boots and gloves.
- Wash your hands thoroughly with soap and clean water often when working in debris.

Safety Issues

- Be aware of new safety issues created by the disaster. Watch for washed out roads, contaminated buildings, contaminated water, gas leaks, broken glass, damaged electrical wiring, and slippery floors.
- Inform local authorities about health and safety issues, including chemical spills, downed power lines, washed out roads, etc.
-

Fujita Pearson Tornado Scale

Category	Category Description	Level of Damage
F-0	Gale Tornado 40 - 72 MPH	Chimneys damaged; branches broken off trees; shallow-rooted trees uprooted; sign boards damaged.
F-1	Moderate Tornado 73 - 112 MPH	Roof surfaces peeled off; mobile homes pushed off foundations or overturned; moving autos pushed off roads.
F-2	Significant Tornado 113 - 157 MPH	Roofs torn off frame houses; mobile homes demolished; box cars pushed over; large trees snapped or uprooted; light-object projectiles generated.
F-3	Severe Tornado 158 - 206 MPH	Roofs and some walls torn off well-constructed houses; trains overturned; most trees in forest uprooted; heavy cars lifted off the ground and thrown.
F-4	Devastating Tornado 207 - 260 MPH	Well-constructed houses leveled; structures with weak foundations relocated; cars thrown and large projectiles generated.
F-5	Incredible Tornado 261 - 318 MPH	Strong frame houses lifted off foundations and carried considerable distance to disintegrate; automobile-sized projectiles hurtle through the air in excess of 100 yards; trees debarked; other incredible phenomena expected.

Winter Storms and Extreme Cold

Dressing for the winter weather

- Wear several layers of loose fitting, lightweight, warm clothing rather than one layer of heavy clothing.
- The outer garments should be tightly woven and water repellent.
- Wear a hat.
- Cover your mouth to protect your lungs.

During a winter storm

- The following are guidelines for what you should do during a winter storm or under conditions of extreme cold:
- Listen to your radio, television, or NOAA Weather Radio for weather reports and emergency information.
- Eat regularly and drink ample fluids, but avoid caffeine and alcohol.
- Avoid overexertion. Overexertion can bring on a heart attack —a major cause of death in the winter.
- When exposed to frigid temperatures and/or stormy conditions:
- Watch for signs of frostbite. These include loss of feeling and white or pale appearance in extremities such as fingers, toes, ear lobes, and the tip of the nose. If symptoms are detected, get medical help immediately.
- Watch for signs of hypothermia. These include uncontrollable shivering, memory loss, disorientation, incoherence, slurred speech, drowsiness, and apparent exhaustion.
- If symptoms of hypothermia are detected, get the victim to a warm location, remove wet clothing, warm the center of the body first, and give warm, non-alcoholic beverages if the victim is conscious. Get medical help as soon as possible.

Winter weather conditions--terms

- Freezing Rain: Rain that freezes when it hits the ground, creating a coating of ice on roads, walkways, trees, and power lines.
- Sleet: Rain that turns to ice pellets before reaching the ground. Sleet also causes moisture on roads to freeze and become slippery.
- Winter Storm Watch: A winter storm is possible in your area. Tune in to NOAA Weather Radio, commercial radio, or television for more information.
- Winter Storm Warning: A winter storm is occurring or will soon occur in your area.
- Blizzard Warning: Sustained winds or frequent gusts to 35 miles per hour or greater and considerable amounts of falling

or blowing snow (reducing visibility to less than a quarter mile) are expected to prevail for a period of three hours or longer.

- Frost/Freeze Warning: Below freezing temperatures are expected.

Wind Chill Table

Little Danger | Increasing Danger | Greater Danger that Exposed Flesh Will Freeze

WIND VELOCITY (mph)

TEMPERATURE (°F)	0	5	10	15	20	25	30	35	40	45	50
-10	-10	-15	-31	-45	-52	-58	-63	-67	-69	-70	-70
-5	-5	-11	-27	-40	-46	-52	-56	-60	-62	-63	-63
0	0	-6	-22	-33	-40	-45	-49	-52	-54	-54	-56
5	5	1	-15	-25	-32	-37	-41	-43	-45	-46	-47
10	10	7	-9	-18	-24	-29	-33	-35	-36	-38	-38
15	15	12	-2	-11	-17	-22	-26	-27	-29	-31	-31
20	20	16	2	-6	-9	-15	-18	-20	-22	-24	-24
25	25	21	9	1	-4	-7	-11	-13	-15	-17	-17
30	30	27	16	11	3	0	-2	-4	-4	-6	-7
35	35	33	21	16	12	7	5	3	1	1	0
40	40	37	28	22	18	16	13	11	10	9	8

Extreme Heat

During a heat emergency

- Stay indoors as much as possible and limit exposure to the sun.
- Stay on the lowest floor out of the sunshine if air conditioning is not available.
- Circulating air can cool the body by increasing the perspiration rate of evaporation.
- Eat well-balanced, light, and regular meals. Avoid using salt tablets unless directed to do so by a physician.
- Drink plenty of water. Persons who have epilepsy or heart, kidney, or liver disease; are on fluid-restricted diets; or have a problem with fluid retention should consult a doctor before increasing liquid intake.
- Limit intake of alcoholic beverages.
- Dress in loose-fitting, lightweight clothes. When outdoors in the sun wear light-colored clothes that cover as much skin as possible.
- Protect face and head.
- Use a buddy system when working in extreme heat,
- and take frequent breaks.

Heat Index Chart

Air Temperature (°F)	Relative Humidity (%)																		
	10	15	20	25	30	35	40	45	50	55	60	65	70	75	80	85	90	95	100
130	131																		
125	123	131	141																
120	116	123	130	139	148														
115	111	115	120	127	135	143	151												
110	105	106	112	117	123	130	137	143	150										
105	100	102	105	109	113	118	123	129	135	142	149								
100	95	97	99	101	104	107	110	115	120	126	132	138	144						
95	90	91	93	94	96	98	101	104	107	110	114	119	124	130	136				
90	85	86	87	88	90	91	93	95	96	98	100	102	106	109	113	117	122		
85	80	81	82	83	84	85	86	87	88	89	90	91	93	95	97	99	102	105	106
80	75	76	77	77	78	79	79	80	81	81	82	83	85	86	86	87	88	89	91
75	70	71	72	72	73	73	74	74	75	75	76	76	77	77	78	78	79	79	80
70	65	65	66	66	67	67	68	68	69	69	70	70	70	70	71	71	71	71	72

Heat Index Chart (Apparent Temperature)

Extreme heat terminology:

- *Heat wave:* Prolonged period of excessive heat often combined with excessive humidity.
- *Heat index:* A number in degrees Fahrenheit (F) that tells how hot it feels when relative humidity is added to the air temperature. Exposure to full sunshine can increase the heat index by 15 degrees.
- *Heat cramps:* Muscular pains and spasms due to heavy exertion. Although heat cramps are the least severe, they are often the first signal that the body is having trouble with the heat.
- *Heat exhaustion:* Typically occurs when people exercise heavily or work in a hot, humid place where body fluids are lost through heavy sweating. Blood flow to the skin increases, causing blood flow to decrease to the vital organs. This results in a form of mild shock. If not treated, the victim's condition will worsen. Body temperature will keep rising and the victim may suffer heat stroke.
- *Heat stroke:* A life-threatening condition. The victim's temperature control system, which produces sweating to cool the body, stops working. The body temperature can rise so high that brain damage and death may result if the body is not cooled quickly.
- *Sun stroke:* Another term for heat stroke.

Earthquakes

During an Earthquake

When indoors

- Minimize your movements to a few steps to a nearby safe place.
- Stay indoors until shaking has stopped and you're sure exiting is safe.
- Take cover under a sturdy desk, table, or bench, or against an inside wall, and hold on.
- If there isn't a table or desk near you, cover your face and head with your arms and crouch in an inside corner of the building.
- Stay away from glass, windows, outside doors and walls, and anything that could fall, such as lighting fixtures or furniture.
- If you are in bed when the earthquake strikes, stay there. Protect your head with a pillow unless you are under or beside heavy items that could fall. In that case, move to the nearest safe place.
- Use a doorway for shelter only if it is in close proximity to you and if you know it is a strongly supported, load-bearing doorway.
- Stay inside until the shaking stops and it is safe to go outside. Most injuries during earthquakes occur when people are hit by falling objects when entering into or exiting from buildings.
- Be aware that the electricity may go out or the sprinkler systems or fire alarms may turn on.
- Do not use the elevators.

When outside

- Minimize your movements to a few steps to get you to a nearby safe place or away from potentially dangerous structures.
- Move away from buildings, streetlights, and utility wires.

In a moving vehicle

- Stop as quickly as safety permits and stay in the vehicle. Avoid stopping near/under buildings, trees, overpasses, and utility wires.
- Proceed cautiously once the earthquake has stopped, watching for road and bridge damage.

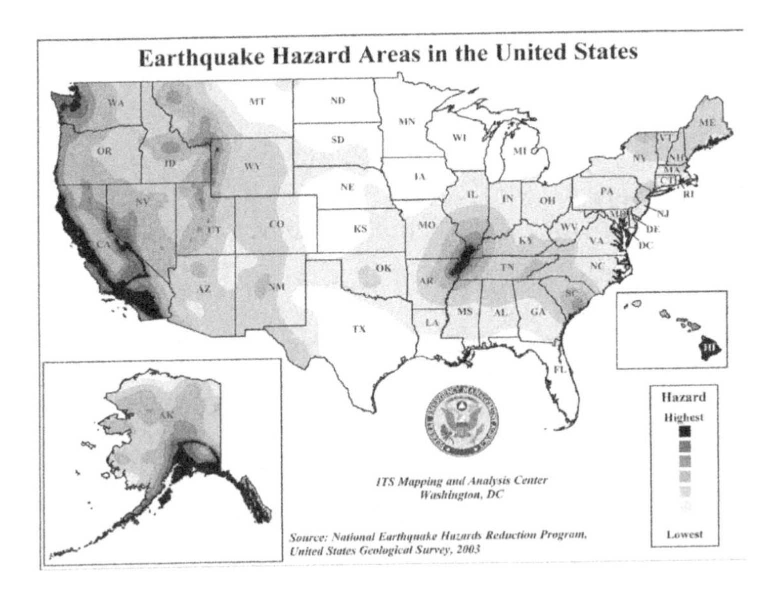

After an Earthquake

- Be prepared for aftershocks. These secondary shockwaves are usually less violent than the main quake but can be strong enough to do additional damage to weakened structures.
- Open cabinets cautiously. Objects may fall off shelves.
- Stay away from damaged areas unless your assistance has been specifically requested by police, fire, or relief organizations.
- Be aware of possible tsunamis if you live in coastal areas. These are also known as seismic sea waves (mistakenly called "tidal waves"). When local authorities issue a tsunami warning, assume

If trapped under debris

- Do not light a match.
- Do not move about or kick up dust.
- Cover your mouth with a handkerchief or clothing.
- Tap on a pipe or wall so rescuers can locate you. Use a whistle if one is available. Shout only as a last resort—shouting can cause you to inhale dangerous amounts of dust.

Earthquake Hazards by State

Earthquake terminology

- *Earthquake:* A sudden slipping or movement of a portion of the earth's crust accompanied and followed by a series of vibrations.
- *Aftershock:* An earthquake of similar or lesser intensity that

follows the main earthquake.

- *Fault:* The fracture across which displacement has occurred during an earthquake. The slippage may range from less than an inch to more than 10 yards in a severe earthquake.
- *Epicenter:* The place on the earth's surface directly above the point on the fault where the earthquake rupture began. Once fault slippage begins, it expands along the fault during the earthquake and can extend hundreds of miles before stopping.
- *Seismic waves:* Vibrations that travel outward from the earthquake fault at speeds of several miles per second. Fault slippage directly under a structure can cause considerable damage, the vibrations of seismic waves cause most of the destruction during earthquakes.
- *Magnitude:* The amount of energy released during an earthquake, which is computed from the amplitude of the seismic waves. A magnitude of 7.0 on the Richter Scale indicates an extremely strong earthquake. Each whole number on the scale represents an increase of about 30 times more energy released than the previous whole number represents. Therefore, an earthquake measuring 6.0 is about 30 times more powerful than one measuring 5.0.

The Modified Mercalli Scale

Scale Range		Level of Damage	Richter Scale
1-4	Instrumental to Moderate	No damage.	</= 4.3
5	Rather Strong	Damage negligible. Small, unstable objects displaced or upset; some dishes and glassware broken.	4.4 - 4.8
6	Strong	Damage slight. Windows, dishes, glassware broken. Furniture moved or overturned. Weak plaster and masonry cracked.	4.9 - 5.4
7	Very Strong	Damage slight-moderate in well-built structures; considerable in poorly-built structures. Furniture and weak chimneys broken. Masonry damaged. Loose bricks, tiles, plaster, and stones will fall.	5.5 - 6.1
8	Destructive	Structure damage considerable, particularly to poorly built structures. Chimneys, monuments, towers, elevated tanks may fail. Frame houses moved. Trees damaged. Cracks in wet ground and steep slopes.	6.2 - 6.5
9	Ruinous	Structural damage severe; some will collapse. General damage to foundations. Serious damage to reservoirs. Underground pipes broken. Conspicuous cracks in ground; liquefaction.	6.6 - 6.9
10	Disastrous	Most masonry and frame structures/foundations destroyed. Some well-built wooden structures and bridges destroyed. Serious damage to dams, dikes, embankments. Sand and mud shifting on beaches and flat land.	7.0 - 7.3
11	Very Disastrous	Few or no masonry structures remain standing. Bridges destroyed. Broad fissures in ground. Underground pipelines completely out of service. Rails bent. Widespread earth slumps and landslides.	7.4 - 8.1
12	Catastrophic	Damage nearly total. Large rock masses displaced. Lines of sight and level distorted.	> 8.1

Volcanoes

During a volcanic eruption
Evacuate immediately from the volcano area to avoid flying debris, hot gases, lateral blast, heavy falling ash, and lava flow.

Gases
Most gases from a volcano quickly blow away. However, heavy gases such as carbon dioxide and hydrogen sulfide can collect in low-lying areas. The most common volcanic gas is water vapor, followed by carbon dioxide and sulfur dioxide. Sulfur dioxide can cause breathing problems in both healthy people and people with asthma and other respiratory problems. Other volcanic gases include hydrogen chloride, carbon monoxide, and hydrogen fluoride. Amounts of these gases vary widely from one volcanic eruption to the next.
Although gases usually blow away rapidly, it is possible that people who are close to the volcano or who are in the low-lying areas downwind will be exposed to levels that may affect health. At low levels, gases can irritate the eyes, nose, and throat. At higher levels, gases can cause rapid breathing, headache, dizziness, swelling and spasm of the throat, and suffocation.

Falling Ash
- Wear long-sleeved shirts and long pants.
- Use goggles and wear eyeglasses instead of contact lenses.
- Use a dust mask or hold a damp cloth over your face to aid breathing.
- Stay away from areas downwind from the volcano to avoid volcanic ash.
- Stay indoors until the ash has settled unless there is danger of the roof collapsing.
- Close doors, windows, and all ventilation in the house (chimney vents, furnaces, air conditioners, fans, and other vents).
- Clear heavy ash from flat or low-pitched roofs and rain gutters.
- Avoid running car or truck engines. Driving can stir up volcanic ash that can clog engines, damage moving parts, and stall vehicles.
- Avoid driving in heavy ash fall unless absolutely required. If you have to drive, keep speed down to 35 MPH or slower.

The Volcanic Explosivity Index

Category	Category Description
0	Non-Explosive (Hawaiian) Plume: < 100 m/Volume: > 1000 m3
1	Gentle (Hawaiian - Strombolian) Plume: 100 - 1000 m/Volume: >10,000 m3
2	Explosive (Strombolian - Vulcanian) Plume: 1 - 5 km/Volume: > 1,000,000 m3
3	Severe (Vulcanian) Plume: 3 - 15 km/Volume: > 10,000,000 m3
4	Cataclysmic (Vulcanian - Plinian) Plume: 10 - 25 km/Volume: > 100,000,000 m3
5	Paroxysmal (Plinian) Plume: >25 km/Volume: > 1 km3
6	Colossal (Plinian - Ultraplinian) Plume: > 25 km/Volume: > 10 km3
7	No Adjectival Description Plume: > 25 km/Volume: > 100 km3
8	No Adjectival Description Plume: > 25 km/Volume: > 1,000 km3

Tsunamis

During a tsunami

- Turn on your radio to learn if there is a tsunami warning if an earthquake occurs and you are in a coastal area.
- Move inland to higher ground immediately and stay there.

WARNING!
If there is noticeable recession in water way from the shoreline this is nature's tsunami warning and it should be heeded. You should move away immediately!

After a tsunami

- Stay away from flooded and damaged areas until officials say it is safe to return.
- Stay away from debris in the water; it may pose a safety hazard to boats and people.

Tsunami terminology

- *Tsunami advisory:* An earthquake has occurred in the Pacific basin, which might generate a tsunami.

- *Tsunami watch:* A tsunami was or may have been generated, but is at least two hours travel time to the area in Watch status.
- *Tsunami warning:* A tsunami was, or may have been generated, which could cause damage; therefore, people in the warned area are strongly advised to evacuate.

Landslides

During a landslide or debris flow

- Stay alert and awake. Many debris-flow fatalities occur when people are sleeping. Listen to a NOAA Weather Radio or portable, battery-powered radio or television for warnings of intense rainfall. Be aware that intense, short bursts of rain may be particularly dangerous, especially after longer periods of heavy rainfall and damp weather.
- If you are in areas susceptible to landslides and debris flows, consider leaving if it is safe to do so. Remember that driving during an intense storm can be hazardous. If you remain at home, move to a second story if possible. Staying out of the path of a landslide or debris flow saves lives.
- Listen for any unusual sounds that might indicate moving debris, such as trees cracking or boulders knocking together. A trickle of flowing or falling mud or debris may precede larger landslides. Moving debris can flow quickly and sometimes without warning.
- If you are near a stream or channel, be alert for any sudden increase or decrease in water flow and for a change from clear to muddy water. Such changes may indicate landslide activity upstream, so be prepared to move quickly. Don't delay! Save yourself, not your belongings.
- Be alert when driving. Embankments along roadsides are particularly susceptible to landslides. Watch the road for collapsed pavement, mud, fallen rocks, and other indications of possible debris flows.

What to do if you suspect imminent landslide danger

- Evacuate. Getting out of the path of a landslide or debris flow is your best protection.
- Curl into a tight ball and protect your head if escape is not possible.

After a landslide or debris flow

- Stay away from the slide area. There may be danger of additional slides.
- Check for injured and trapped persons near the slide, without

entering the direct slide area. Direct rescuers to their locations.
- Watch for associated dangers such as broken electrical, water, gas, and sewage lines and damaged roadways and railways.

Wildfires

During a wildfire

- Gather fire tools such as a rake, axe, handsaw or chainsaw, bucket, and shovel.
- Park your vehicle it in an open space facing the direction of escape. Shut doors and roll up windows. Leave the key in the ignition and the doors unlocked.
- If advised to evacuate, do so immediately. Choose a route away from the fire hazard. Watch for changes in the speed and direction of the fire and smoke.

Notes:

TERRORISM

! Disclaimer: this information is advisory in nature and is not intended to identify all scenarios or situations a person might encounter.

! Following these guidelines will not guarantee your safety in every situation.

Living with the threat

We live with many dangers in our daily lives, ranging from everyday accidents to natural disasters. We do so without relentless fear. Terrorism is a fact of contemporary life, but we do not have to live in constant fear of terrorism any more than other dangers. It is important to be aware of the threat and take steps to protect ourselves, but it is also important to keep the threat in perspective.

While there is no absolute protection against terrorism, there are a number of reasonable precautions that can provide some degree of individual protection.

Your biggest asset is the ability to learn how to maintain a state of awareness relative to your environment, and prepare to react if a terrorist event occurs.. Also, don't ever discount what seems to be a sixth sense or premonition. They are sometimes the result of inadvertently picking up cues from your environment. It is better to take precautions and perhaps risk a little embarrassment than to disregard what might be lifesaving feelings.

Weapons of Mass Destruction (WMD's)

A weapon of mass destruction (WMD) is any weapon that can kill large numbers of humans and can also cause great damage to man-made and natural structures.

WMD's are categorized as the following:

CBRNE

- **C**hemical **B**iological **R**adiological **N**uclear **E**xplosive

Chemical Terrorism

Chemical Emergencies

A chemical emergency occurs when a hazardous chemical has been released and the release has the potential for harming people's health. Chemical releases can be unintentional, as in the case of an industrial accident or intentional, as in the case of a terrorist attack.

Where Hazardous Chemicals Come From

Some chemicals that are hazardous have been developed by military organizations for use in warfare. Examples are nerve agents such as sarin and VX, mustards such as sulfur mustards and nitrogen mustards, and choking agents such as phosgene. It might be possible for terrorists to get these chemical warfare agents and use them to harm people.

Many hazardous chemicals are used in industry (for example, chlorine, ammonia, and benzene). Others are found in nature (for example, poisonous plants). Some could be made from everyday items such as household cleaners. These types of hazardous chemicals also could be obtained and used to harm people, or they could be accidentally released.

Types and Categories of Hazardous Chemicals

Scientists often categorize hazardous chemicals by the type of chemical or by the effects a chemical would have on people exposed to it. The categories/types used by the Centers for Disease Control and Prevention are as follows:

- Biotoxins poisons that come from plants or animals
- Blister agents/vesicants chemicals that severely blister the eyes, respiratory tract, and skin on contact
- Blood agents poisons that affect the body by being absorbed into the blood
- Caustics (acids) chemicals that burn or corrode people's skin, eyes, and mucus membranes (lining of the nose, mouth, throat, and lungs) on contact
- Choking/lung/pulmonary agents chemicals that cause severe irritation or swelling of the respiratory tract (lining of the nose and throat, lungs)
- Incapacitating agents drugs that make people unable to think clearly or that cause an altered state of consciousness (possibly unconsciousness)
- Long-acting anticoagulants poisons that prevent blood from clotting properly, which can lead to uncontrolled bleeding
- Metals agents that consist of metallic poisons
- Nerve agents highly poisonous chemicals that work by preventing the nervous system from working properly
- Organic solvents agents that damage the tissues of living things by dissolving fats and oils
- Riot control agents/tear gas highly irritating agents normally

used by law enforcement for crowd control or by individuals for protection (for example, mace)

- Toxic alcohols poisonous alcohols that can damage the heart, kidneys, and nervous system
- Vomiting agents chemicals that cause nausea and vomiting

Chemical Agents

Name	Color	Smell
Nerve Agents		
Tabun (GA)	Colorless to brown	Fruity
Sarin (GB)	Colorless	No odor
Soman (GD)	Colorless	Fruity; oil of camphor
VX	Colorless to straw color	No odor
Vesicants		
Impure sulfur mustard (H)	Pale yellow to dark brown	Garlic or mustard
Distilled sulfur mustard (HD)	Pale yellow to dark brown	Garlic or mustard
Lewisite (L)	Pure: colorless Agent: amber to dark brown	Geranium
Riot Control Agents		
Chlorobenzylidene Malononitrile (CS)	White crystalline powder	Pepper
Chloroacetophenon e (CN)	Liquid or solid	Apple blossom
Diphenylaminearsi ne (DM)	Yellow-green crystalline solid	No odor
Pulmonary Agents		
Chlorine (Cl2)	Clear to yellow gas	Bleach
Phosgene (CG)	Colorless gas	Freshly-mown hay
Cyanides (Blood Agents)		
Hydrogen Cyanide (AC)	Gas	Bitter almonds or peach kernels
Cyanogen Chloride	Gas or liquid—colorless	Pungent, biting odor
Incapacitating Agents		
BZ	White crystalline powder	No odor

Chemical Agents: Facts About Evacuation

Some kinds of chemical accidents or attacks, such as a train derailment or a terrorist incident, may make staying put dangerous. In such cases, it may be safer for you to evacuate, or leave the immediate area. You may need to go to an emergency shelter after you leave the immediate area.

How to know if you need to evacuate

- You will hear from the local police, emergency coordinators, or government on the radio and/or television emergency broadcast system if you need to evacuate.
- If there is a "code red" or "severe" terror alert, you should pay attention to radio and/or television broadcasts so you will know right away if an evacuation order is made for your area.
- Every emergency is different and during any emergency people may have to evacuate or to shelter in place depending on where they live.

What to evacuate

- Move away immediately in a direction upwind of the source
- Act quickly and follow the instructions of local emergency coordinators, such as law enforcement personnel, fire departments, or local elected leaders. Every situation can be different, so local coordinators could give you special instructions to follow for a particular situation.
- Local emergency coordinators may direct people to evacuate homes or offices and go to an emergency shelter. If so, emergency coordinators will tell you how to get to the shelter. If you have children in school, they may be sheltered at the school. You should not try to get to the school if the children are being sheltered there. Transporting them from the school will put them, and you, at increased risk.
- The emergency shelter will have most supplies that people need. The emergency coordinators will tell you which supplies to bring with you, but you may also want to prepare a portable supply kit. Be sure to bring any medications you are taking.
- If you have time, call a friend or relative in another state to tell them where you are going and that you are safe. Local telephone lines may be jammed in an emergency, so you should plan ahead to have an out-of-state contact with whom to leave messages.
- If you do not have private transportation, make plans in advance of an emergency to identify people who can give you a ride.
- Evacuating and sheltering in this way should keep you safer than if you stayed at home or at your workplace. You will most likely not be in the shelter for more than a few hours. Emergency coordinators will let you know when it is safe to

leave the shelter and anything you may need to do to make sure it is safe to re-enter your home.

Chemical Agents: Facts About Sheltering In Place

What "sheltering in place" means

Some kinds of chemical accidents or attacks may make going outdoors dangerous. Leaving the area might take too long or put you in harm's way. In such a case it may be safer for you to stay indoors than to go outside.

"Shelter in place" means to make a shelter out of the place you are in. It is a way for you to make the building as safe as possible to protect yourself until help arrives. You should not try to shelter in a vehicle unless you have no other choice. Vehicles are not airtight enough to give you adequate protection from chemicals.

Every emergency is different and during any emergency people may have to evacuate or to shelter in place depending on where they live.

How to prepare to shelter in place

Choose a room in your house or apartment for the shelter. The best room to use for the shelter is a room with as few windows and doors as possible. A large room with a water supply is best something like a master bedroom that is connected to a bathroom. For most chemical events, this room should be as high in the structure as possible to avoid vapors (gases) that sink. This guideline is different from the sheltering-in-place technique used in tornadoes and other severe weather and for nuclear or radiological events, when the shelter should be low in the home.

You might not be at home if the need to shelter in place ever arises, but if you are at home, the following items, many of which you may already have, would be good to have in your shelter room:

- First aid kit
- Flashlight, battery-powered radio, and extra batteries for both
- A working telephone
- Food and bottled water. Store 1 gallon of water per person in plastic bottles as well as ready-to-eat foods that will keep without refrigeration in the shelter-in-place room. If you do not have bottled water, or if you run out, you can drink water from a toilet tank (not from a toilet bowl). Do not drink water from the tap.
- Duct tape and scissors.
- Towels and plastic sheeting. You may wish to cut your plastic sheeting to fit your windows and doors before any emergency

occurs.

How to know if you need to shelter in place

- Most likely you will only need to shelter for a few hours.
- You will hear from the local police, emergency coordinators, or government on the radio and on television emergency broadcast system if you need to shelter in place.
- If there is a "code red" or "severe" terror alert, you should pay attention to radio and television broadcasts to know right away whether a shelter-in-place alert is announced for your area.

What to do

Act quickly and follow the instructions of your local emergency coordinators such as law enforcement personnel, fire departments, or local elected leaders. Every situation can be different, so local emergency coordinators might have special instructions for you to follow. In general, do the following:

- Go inside as quickly as possible. Bring any pets indoors.
- If there is time, shut and lock all outside doors and windows. Locking them may pull the door or window tighter and make a better seal against the chemical.
- Turn off the air conditioner or heater. Turn off all fans, too.
- Close the fireplace damper and any other place that air can come in from outside.
- Go in the shelter-in-place room and shut the door.
- Turn on the radio. Keep a telephone close at hand, but don't use it unless there is a serious emergency.
- Sink and toilet drain traps should have water in them (you can use the sink and toilet as you normally would). If it is necessary to drink water, drink stored water, not water from the tap.

Dead animals/birds/fish	Not just an occasional incident, but numerous animals (wild and domestic, small and large), birds, and fish in the same area
Lack of insect life	Normal insect activity (ground, air, and/or water) missing, dead insects evident in the ground/water surface/shoreline
Physical symptoms	Numerous individuals experiencing unexplained water-like blisters, wheals (similar to bee stings), pinpointed pupils, choking, respiratory ailments and/or rashes
Mass casualties	Numerous individuals exhibiting unexplained serious health problems ranging from nausea to disorientation to difficulty in breathing to convulsions and death
Definite pattern of casualties	Casualties distributed in a pattern that may be associated with possible agent dissemination methods
Illness associated with confined geographic area	Lower incidence of symptoms for people working indoors than outdoors, or the reverse
Unusual liquid droplets	Numerous surfaces exhibiting oily droplets/film; numerous water surfaces displaying an oily film (no recent rain)
Areas that look different in appearance	Not just a patch of dead weeds, but trees, shrubs, bushes, food crops, and/or lawns that are dead, discolored, or withered (no current drought)
Unexplained odors	Smells ranging from fruity to flowery to sharp/pungent to garlic/horseradish-like to bitter almonds/peach kernels to newly mown hay; the particular odor is completely out of character with its surroundings
Low-lying clouds	Low-lying cloud/fog-like condition that is not explained by its surroundings
Unusual metal debris	Unexplained bomb/munitions-like material, especially if it contains a liquid (no recent rain)

Indicators of a Possible Chemical Incident

- Tape plastic over any windows in the room. Use duct tape around the windows and doors and make an unbroken seal. Use the tape over any vents into the room and seal any electrical outlets or other openings.

- Listen to the radio for an announcement indicating that it is safe to leave the shelter.
- When you leave the shelter, follow instructions from local emergency coordinators to avoid any contaminants outside. After you come out of the shelter, emergency coordinators may have additional instructions on how to make the rest of the building safe again.
- If you are away from your shelter-in-place location when a chemical event occurs, follow the instructions of emergency coordinators to find the nearest shelter. If your children are at school, they will be sheltered there. Unless you are instructed to do so, do not try to get to the school to bring your children home. Transporting them from the school will put them, and you, at increased risk.

After a Chemical Attack

Decontamination is needed within minutes of exposure to minimize health consequences. Do not leave the safety of a shelter to go outdoors to help others until authorities announce it is safe to do so.

A person affected by a chemical agent requires immediate medical attention from a professional. If medical help is not immediately available, decontaminate yourself and assist in decontaminating others.

Decontamination Guidelines Are As Follows:

- Use extreme caution when helping others who have been exposed to chemical agents.
- Remove all clothing and other items in contact with the body. Contaminated clothing normally removed over the head should be cut off to avoid contact with the eyes, nose, and mouth. Put contaminated clothing and items into a plastic bag and seal it. Decontaminate hands using soap and water. Remove eyeglasses or contact lenses. Put glasses in a pan of household bleach to decontaminate them, and then rinse and dry.
- Flush eyes with water.
- Gently wash face and hair with soap and water before thoroughly rinsing with water.
- Decontaminate other body areas likely to have been contaminated. Blot (do not swab or scrape) with a cloth soaked in soapy water and rinse with clear water.
- Change into uncontaminated clothes. Clothing stored in drawers or closets is likely to be uncontaminated.
- Proceed to a medical facility for screening and professional treatment.

Biological Terrorism

What Is Bioterrorism?

A bioterrorism attack is the deliberate release of viruses, bacteria, or other germs (agents) used to cause illness or death in people, animals, or plants. These agents are typically found in nature, but it is possible that they could be changed to increase their ability to cause disease, make them resistant to current medicines, or to increase their ability to be spread into the environment. Terrorists may use biological agents because they can be extremely difficult to detect and do not cause illness for several hours to several days.

Biological agents can be spread through the air, through water, or in food. Some bioterrorism agents, like the smallpox virus, can be spread from person to person and some, like anthrax, cannot.

Bioterrorism Agent Categories

Bioterrorism agents can be separated into three categories, depending on how easily they can be spread and the severity of illness or death they cause. Category A agents are considered the highest risk and Category C agents are those that are considered emerging threats for disease.

Category A

- These high-priority agents include organisms or toxins that pose the highest risk to the public and national security because:
- They can be easily spread or transmitted from person to person
- They result in high death rates and have the potential for major public health impact
- They might cause public panic and social disruption
- They require special action for public health preparedness.

Category B

- These agents are the second highest priority because:
- They are moderately easy to spread
- They result in moderate illness rates and low death rates
- They require specific enhancements of CDC's laboratory capacity and enhanced disease monitoring.

Category C

- These third highest priority agents include emerging pathogens that could be engineered for mass spread in the future because:
- They are easily available
- They are easily produced and spread
- They have potential for high morbidity and mortality rates and major health impact.

Bombs

Radiological (Dirty Bombs)

People have expressed concern about dirty bombs and what they should do to protect themselves if a dirty bomb incident occurs. Because your health and safety are our highest priorities, the health experts at the Centers for Disease Control and Prevention (CDC) have prepared the following list of frequently asked questions and answers about dirty bombs.

What is a dirty bomb?

A dirty bomb is a mix of explosives, such as dynamite, with radioactive powder or pellets. When the dynamite or other explosives are set off, the blast carries radioactive material into the surrounding area.

A dirty bomb is not the same as an atomic bomb. An atomic bomb, like those bombs dropped on Hiroshima and Nagasaki, involves the splitting of atoms and a huge release of energy that produces the atomic mushroom cloud. A dirty bomb works completely differently and cannot create an atomic blast. Instead, a dirty bomb uses dynamite or other explosives to scatter radioactive dust, smoke, or other material in order to cause radioactive contamination.

What are the main dangers of a dirty bomb?

The main danger from a dirty bomb is from the explosion, which can cause serious injuries and property damage. The radioactive materials used in a dirty bomb would probably not create enough radiation exposure to cause immediate serious illness, except to those people who are very close to the blast site. However, the radioactive dust and smoke spread farther away could be dangerous to health if it is inhaled. Because people cannot see, smell, feel, or taste radiation, you should take immediate steps to protect yourself and your loved ones.

Protecting Yourself

These simple steps recommended by doctors and radiation experts will help protect you and your loved ones. The steps you should take depend on where you are located when the incident occurs: outside, inside, or in a vehicle.

If you are outside and close to the incident

- Cover your nose and mouth with a cloth to reduce the risk of breathing in radioactive dust or smoke.
- Don't touch objects thrown off by an explosion they might be radioactive.
- Quickly go into a building where the walls and windows have not been broken. This area will shield you from radiation that might be outside.
- Once you are inside, take off your outer layer of clothing and seal it in a plastic bag if available. Put the cloth you used to cover your mouth in the bag, too. Removing outer clothes may get rid of up to 90% of radioactive dust.

- Put the plastic bag where others will not touch it and keep it until authorities tell you what to do with it.
- Shower or wash with soap and water. Be sure to wash your hair. Washing will remove any remaining dust.
- Tune to the local radio or television news for more instructions.

If you are inside and close to the incident

- If the walls and windows of the building are not broken, stay in the building and do not leave.
- To keep radioactive dust or powder from getting inside, shut all windows, outside doors, and fireplace dampers. Turn off fans and heating and air-conditioning systems that bring in air from the outside. It is not necessary to put duct tape or plastic around doors or windows.
- If the walls and windows of the building are broken, go to an interior room and do not leave. If the building has been heavily damaged, quickly go into a building where the walls and windows have not been broken.
- If you must go outside, be sure to cover your nose and mouth with a cloth. Once you are back inside, take off your outer layer of clothing and seal it in a plastic bag if available. Store the bag where others will not touch it.
- Shower or wash with soap and water, removing any remaining dust. Be sure to wash your hair.
- Tune to local radio or television news for more instructions.

If you are in a car when the incident happens

- Close the windows and turn off the air conditioner, heater, and vents.
- Cover your nose and mouth with a cloth to avoid breathing radioactive dust or smoke.
- If you are close to your home, office, or a public building, go there immediately and go inside quickly.
- If you cannot get to your home or another building safely, pull over to the side of the road and stop in the safest place possible. If it is a hot or sunny day, try to stop under a bridge or in a shady spot.
- Turn off the engine and listen to the radio for instructions.
- Stay in the car until you are told it is safe to get back on the road.

What should I do about my children and family?

- If your children or family are with you, stay together. Take the same actions to protect your whole family.
- If your children or family are in another home or building, they should stay there until you are told it is safe to travel.
- Schools have emergency plans and shelters. If your children are at school, they should stay there until it is safe to travel. Do not go to the school until public officials say it is safe to travel.

Should I take potassium iodide?
Potassium iodide, also called KI, only protects a person's thyroid gland from exposure to radioactive iodine. KI will not protect a person from other radioactive materials or protect other parts of the body from exposure to radiation. Since there is no way to know at the time of the explosion whether radioactive iodine was used in the explosive device, taking KI would probably not be beneficial. Also, KI can be dangerous to some people.

Will food and water supplies be safe?
Food and water supplies most likely will remain safe. However, any unpackaged food or water that was out in the open and close to the incident may have radioactive dust on it. Therefore, do not consume water or food that was out in the open. The food inside of cans and other sealed containers will be safe to eat. Wash the outside of the container before opening it. Authorities will monitor food and water quality for safety and keep the public informed.

How do I know if I've been exposed to radiation or contaminated by radioactive materials?
People cannot see, smell, feel, or taste radiation; so you may not know whether you have been exposed. Police or firefighters will quickly check for radiation by using special equipment to determine how much radiation is present and whether it poses any danger in your area.

Low levels of radiation exposure (like those expected from a dirty bomb situation) do not cause any symptoms. Higher levels of radiation exposure may produce symptoms, such as nausea, vomiting, diarrhea, and swelling and redness of the skin. If you develop any of these symptoms, you should contact your doctor, hospital, or other sites recommended by authorities.

Clothing
Some kinds of chemical accidents or attacks may cause you to come in contact with dangerous chemicals. Coming in contact with a dangerous chemical may make it necessary for you to remove and dispose of your clothing right away and then wash yourself. Removing your clothing and washing your body will reduce or remove the chemical so that it is no longer a hazard. This process is called decontamination.

People are decontaminated for two primary reasons:

1. To prevent the chemical from being further absorbed by their body or from spreading on their body, and
2. To prevent the chemical from spreading to other people, including medical personnel, who must handle or who might come in contact with the person who is contaminated with the chemical.

Most chemical agents can penetrate clothing and are absorbed rapidly through the skin. Therefore, the most important and most effective decontamination for any chemical exposure is decontamination done within the first minute or two after exposure.

How to know if you need to wash yourself and dispose of clothing

In most cases, emergency coordinators will let you know if a dangerous chemical has been released and will tell you what to do.

In general, exposure to a chemical in its liquid or solid form will require you to remove your clothing and then thoroughly wash your exposed skin. Exposure to a chemical in its vapor (gas) form generally requires you only to remove clothing and the source of the toxic vapor.

If you think you have been exposed to a chemical release, but you have not heard from emergency coordinators, you can follow the washing and clothing disposal advice in the next section.

If your clothes are contaminated

Act quickly and follow the instructions of local emergency coordinators. Every situation can be different, so local emergency coordinators might have special instructions for you to follow. The most important things to do if you think you may have been exposed to a dangerous chemical are to quickly remove your clothing, wash yourself, and dispose of your clothing:

Removing your clothing:

- Quickly take off clothing that has a chemical on it. Any clothing that has to be pulled over your head should be cut off instead of being pulled over your head.
- If you are helping other people remove their clothing, try to avoid touching any contaminated areas, and remove the clothing as quickly as possible.
- As quickly as possible, wash any chemicals from your skin with large amounts of soap and water. Washing with soap and water will help protect you from any chemicals on your body.
- If your eyes are burning or your vision is blurred, rinse your eyes with plain water for 10 to 15 minutes. If you wear contacts, remove them and put them with the contaminated clothing. Do not put the contacts back in your eyes (even if they are not disposable contacts). If you wear eyeglasses, wash them with soap and water. You can put your eyeglasses back on after you wash them.

Disposing of your clothes:

- After you have washed yourself, place your clothing inside a plastic bag. Avoid touching contaminated areas of the

clothing. If you can't avoid touching contaminated areas, or you aren't sure where the contaminated areas are, wear rubber gloves or put the clothing in the bag using tongs, tool handles, sticks, or similar objects.
- Anything that touches the contaminated clothing should also be placed in the bag. If you wear contacts, put them in the plastic bag too.
- Seal the bag, and then seal that bag inside another plastic bag. Disposing of your clothing in this way will help protect you and other people from any chemicals that might be on your clothes.
- When the local or state health department or emergency personnel arrive, tell them what you did with your clothes. The health department or emergency personnel will arrange for further disposal. Do not handle the plastic bags yourself.
- After you have removed and disposed of your clothing, and washed yourself, you should dress in clothing that is not contaminated. Clothing that has been stored in drawers or closets are unlikely to be contaminated, so it would be a good choice for you to wear.
- You should avoid coming in contact with other people who may have been exposed but who have not yet changed their clothes or washed. Move away from the area where the chemical was released when emergency coordinators tell you to do so.

Nuclear

Frequently Asked Questions About a Nuclear Blast

With the recent threats of terrorism, many people have expressed concern about the likelihood and effects of a nuclear blast. The Centers for Disease Control (CDC) has developed a fact sheet to describe what happens when a nuclear blast occurs, the possible health effects, and what you can do to protect yourself in this type of emergency.

What is a nuclear blast?

A nuclear blast, produced by explosion of a nuclear bomb (sometimes called a nuclear detonation), involves the joining or splitting of atoms (called fusion and fission) to produce an intense pulse or wave of heat, light, air pressure, and radiation. The bombs dropped on Hiroshima and Nagasaki, Japan, at the end of World War II produced nuclear blasts.

When a nuclear device is exploded, a large fireball is created. Everything inside of this fireball vaporizes, including soil and water, and is carried upwards. This creates the mushroom cloud that we associate with a nuclear blast, detonation, or explosion. Radioactive material from the nuclear device mixes with the vaporized material in the mushroom cloud.

As this vaporized radioactive material cools, it becomes condensed and forms particles, such as dust. The condensed radioactive material then falls back to the earth; this is what is known as fallout. Because fallout is in the form of particles, it can be carried long distances on wind currents and end up miles from the site of the explosion. Fallout is radioactive and can cause contamination of anything on which it lands, including food and water supplies.

What are the effects of a nuclear blast?

The effects on a person from a nuclear blast will depend on the size of the bomb and the distance the person is from the explosion. However, a nuclear blast would likely cause great destruction, death, and injury, and have a wide area of impact. In a nuclear blast, injury or death may occur as a result of the blast itself or as a result of debris thrown from the blast. People may experience moderate to severe skin burns, depending on their distance from the blast site. Those who look directly at the blast could experience eye damage ranging from temporary blindness to severe burns on the retina. Individuals near the blast site would be exposed to high levels of radiation and could develop symptoms of radiation sickness (called acute radiation syndrome, or ARS

While severe burns would appear in minutes, other health effects might take days or weeks to appear. These effects range from mild, such as skin reddening, to severe effects such as cancer and death, depending on the amount of radiation absorbed by the body (the dose), the type of radiation, the route of exposure, and the length of time of the exposure.

People may experience two types of exposure from radioactive materials from a nuclear blast: external exposure and internal exposure. External exposure would occur when people were exposed to radiation outside of their bodies from the blast or its fallout. Internal exposure would occur when people ate food or breathed air that was contaminated with radioactive fallout. Both internal and external exposure from fallout could occur miles away from the blast site. Exposure to very large doses of external radiation may cause death within a few days or months. External exposure to lower doses of radiation and internal exposure from breathing or eating food contaminated with radioactive fallout may lead to an increased risk of developing cancer and other health effects.

How can I protect my family and myself during a nuclear blast?

In the event of a nuclear blast, a national emergency response plan would be activated and would include federal, state, and local agencies. Following are some steps recommended by the World Health Organization if a nuclear blast occurs:

If you are near the blast when it occurs:

- Turn away, close and cover your eyes to prevent damage to your sight.
- Drop to the ground face down and place your hands under your body.
- Remain flat until the heat and two shock waves have passed.

If you are outside when the blast occurs:

- Find something to cover your mouth and nose, such as a scarf, handkerchief, or other cloth.
- Remove any dust from your clothes by brushing, shaking, and wiping in a ventilated area however, cover your mouth and nose while you do this.
- Move to a shelter, basement, or other underground area, preferably located away from the direction that the wind is blowing.
- Remove clothing since it may be contaminated; if possible, take a shower, wash your hair, and change clothes before you enter the shelter.

If you are already in a shelter or basement:

- Cover your mouth and nose with a face mask or other material (such as a scarf or handkerchief) until the fallout cloud has passed.
- Shut off ventilation systems and seal doors or windows until the fallout cloud has passed. However, after the fallout cloud has passed, unseal the doors and windows to allow some air circulation.
- Stay inside until authorities say it is safe to come out.
- Listen to the local radio or television for information and advice. Authorities may direct you to stay in your shelter or evacuate to a safer place away from the area.
- If you must go out, cover your mouth and nose with a damp towel.
- Use stored food and drinking water. Do not eat local fresh food or drink water from open water supplies.
- Clean and cover any open wounds on your body.

If you are advised to evacuate:

- Listen to the radio or television for information about evacuation routes, temporary shelters, and procedures to follow.
- Before you leave, close and lock windows and doors and turn off air conditioning, vents, fans, and furnace. Close fireplace dampers.
- Take disaster supplies with you (such as a flashlight and extra batteries, battery-operated radio, first aid kit and manual, emergency food and water, nonelectric can opener, essential medicines, cash and credit cards, and sturdy shoes).
- Remember your neighbors may require special assistance, especially infants, elderly people, and people with disabilities.

Is a nuclear bomb the same as a suitcase bomb?
The "suitcase" bombs that have been described in new stories in recent years are small nuclear bombs. A suitcase bomb would produce a nuclear blast that is very destructive, but not as great as a nuclear weapon developed for strategic military purposes.
Is a nuclear bomb the same as a dirty bomb?
A nuclear blast is different than a dirty bomb. A dirty bomb, or radiological dispersion device, is a bomb that uses conventional explosives such as dynamite to spread radioactive materials in the form of powder or pellets. It does not involve the splitting of atoms to produce the tremendous force and destruction of a nuclear blast, but rather spreads smaller amounts radioactive material into the surrounding area. The main purpose of a dirty bomb is to frighten people and contaminate buildings or land with radioactive material.
Would an airplane crash in a nuclear power plant have the same effect as a nuclear blast?
While a serious event such as a plane crash into a nuclear power plant could result in a release of radioactive material into the air, a nuclear power plant would not explode like a nuclear weapon. There may be a radiation danger in the surrounding areas, depending on the type of incident, the amount of radiation released, and the current weather patterns. However, radiation would be monitored to determine the potential danger, and people in the local area would be evacuated or advised on how to protect themselves.

BOMBING EVENT

! Disclaimer: this information is advisory in nature and is not intended to identify all scenarios or situations a person might encounter.

! Following these guidelines will not guarantee your safety.

During a Bombing

If there is an explosion, you should immediately:

- Get under a sturdy table or desk if things are falling around you. When things stop falling, leave quickly, watching for obviously weakened floors and stairways.

As you exit the building, be especially watchful of falling debris.
Leave the building as quickly as possible. Do not stop to retrieve personal possessions or make phone calls.
Do not use elevators.

	*WARNING! Secondary devices are always a possibility. A common tactic is to detonate a device attracting a crowd then detonate a second device to inflict heavy casualties.*

What you should do after a bombing

- Leave the area immediately.

Avoid crowds. Crowds of people may be targeted for a second attack.
Avoid unattended cars and trucks. Unattended cars and trucks may contain explosives.
Stay away from damaged buildings to avoid falling glass and bricks. Move at least 10 blocks or 200 yards away from damaged buildings.
Call 9-1-1 once you are in a safe area, but only if police, fire, or EMS has not arrived.
Follow directions from people in authority (police, fire, EMS, or military personnel, or from school or workplace supervisors).
Help others who are hurt or need assistance to leave the area if you are able. If you see someone who is seriously injured, seek help. Do not try to manage the situation alone.
Leave and stay away from the scene of the event. Returning to the scene will increase the risk of danger for rescue workers and you.

9-1-1 services (police, fire, EMS and ambulance) might be delayed indefinitely following a terrorist event, therefore:

- If rescue workers are not available to transport you or other injured persons, always have a back-up plan for

transportation.
Follow advice from your local public safety offices (local health department, local emergency management offices, fire and police departments and reliable news sources).
Triage. Following a terrorist attack or other disasters, injuries are generally treated on a "worst first" basis, called "triage." Triage is not "first come, first served". If your injuries are not immediately life threatening, others might be treated before you. The goal of triage is to save as many lives as possible.
Listen to your radio or television for news and instructions.

Blast Injuries

The four basic mechanisms of blast injury are termed as primary, secondary, tertiary, and quaternary
"Blast Wave" (primary) refers to the intense over-pressurization impulse created by a detonated HE. Blast injuries are characterized by anatomical and physiological changes from the direct or reflective over-pressurization force impacting the body's surface. The HE "blast wave" (over-pressure component) should be distinguished from "blast wind" (forced super-heated air flow). The latter may be encountered with both HE and LE.

When to Go To the Hospital or Clinic

For First Aid, see "Blast Injury" in first aid section

- Seek medical attention if you have any injuries including the following:

Excessive bleeding
Trouble breathing
Persistent cough
Trouble walking or using an arm or leg
Stomach, back or chest pains
Headache
Blurred vision or burning eyes
Dry mouth
Vomiting or diarrhea
Rash or burning skin
Hearing problems
Injuries that increase in pain, redness or swelling
Injuries that do not improve after 24 to 48 hours

Where to Go For Care

Go to a hospital or clinic away from the event if you can. Most victims will go to the nearest hospital. Hospitals away from the event will be less busy.
Expect long waits. To avoid long waits, choose a hospital farther away from the event. While this might increase your travel time, you might receive care sooner.

Limited information. In a large-scale emergency such as a terrorist attack, police, fire, EMS, and even hospitals and clinics cannot track every individual by name. Keep in mind that it may be difficult for hospitals to provide information about loved ones following a terrorist attack. Be patient as you seek information.

Classification of Explosives

Explosives are categorized as high-order explosives (HE) or low-order explosives (LE). HE produce a defining supersonic over-pressurization shock wave. Examples of HE include TNT, C-4, Semtex, nitroglycerin, dynamite, and ammonium nitrate fuel oil (ANFO). LE create a subsonic explosion and lack HE's over-pressurization wave. Examples of LE include pipe bombs, gunpowder, and most pure petroleum-based bombs such as Molotov cocktails or aircraft improvised as guided missiles. HE and LE cause different injury patterns.

Explosive and incendiary (fire) bombs are further characterized based on their source. "Manufactured" implies standard military-issued, mass produced, and quality-tested weapons. "Improvised" describes weapons produced in small quantities, or use of a device outside its intended purpose, such as converting a commercial aircraft into a guided missile. Manufactured (military) explosive weapons are exclusively HE-based. Terrorists will use whatever is available - illegally obtained manufactured weapons or improvised explosive devices (also known as "IEDs") that may be composed of HE, LE, or both. Manufactured and improvised bombs cause markedly different injuries.

How to Recognize and Handle a Suspicious Package or Envelope

	*WARNING!* *If a package or envelope appears suspicious, do not open it.*

Postal Service Mail

There is a long list of possible indicators; these are some of the most common:

- The package or letter has no postage, non-cancelled postage, excessive postage, has been hand delivered, or dropped off by a friend

Sender is unknown or no return address available
Addressee does not normally receive mail at that address
Common words are misspelled
Package emits a peculiar or suspicious odor

Letter or package seems heavy or bulky for its size Package makes a ticking, buzzing, or whirring noise
An unidentified person calls to ask if the letter or package was received

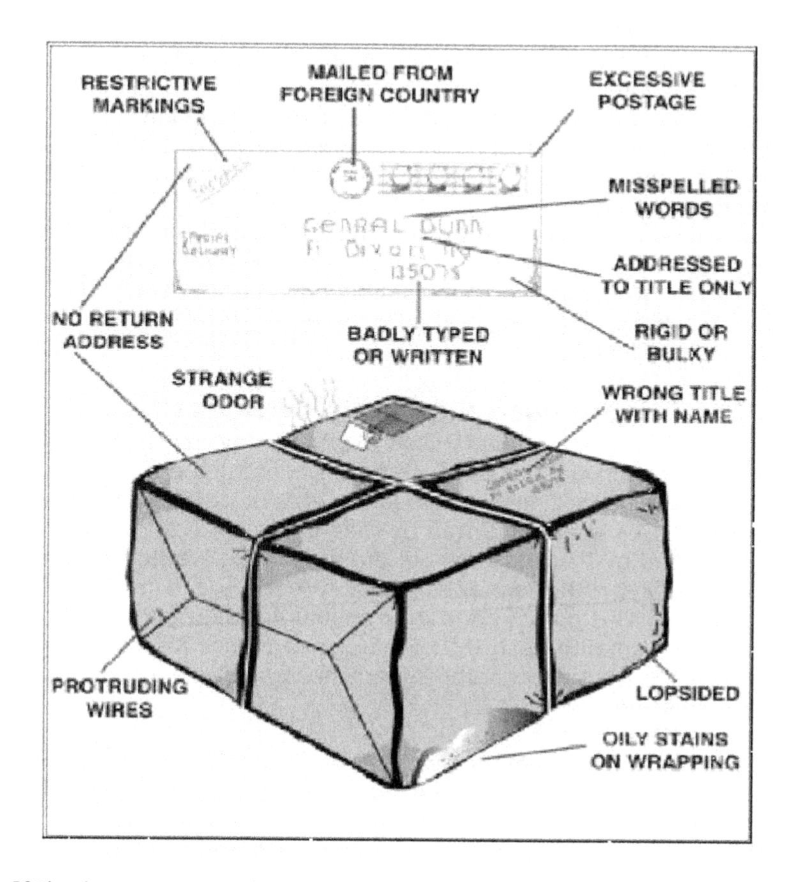

If the letter or parcel exhibits some of the indicators above, it could be considered suspect and the proper authorities should be notified. Never accept unexpected packages at your home, and make sure family members and clerical staff refuse unexpected mail.

Handling of Suspicious Packages or Envelopes

- Do not shake or empty the contents of any suspicious package or envelope.

Do not carry the package or envelope, show it to others or allow others to examine it.
Put the package or envelope down on a stable surface; do not sniff, touch, taste, or look closely at it or at any contents which may have spilled.
Alert others in the area about the suspicious package or envelope.
Leave the area, close any doors, and take actions to prevent others from entering the area. If possible, shut off the ventilation system.
Wash hands with soap and water to prevent spreading potentially infectious material to face or skin. Seek additional instructions for exposed or potentially exposed persons.
If at work, notify a supervisor, a security officer, or a law enforcement official. If at home, contact the local law enforcement agency.
If possible, create a list of persons who were in the room or area when this suspicious letter or package was recognized and a list of persons who also may have handled this package or letter. Give this list to both the local public health authorities and law enforcement officials

IED's (Improvised Explosive Devices)

This is an overview of improvised explosive devices (IEDs). IED is a term for an explosive device that is constructed in an improvised manner designed to kill, maim, or destroy property. These devices are categorized by their container (i.e., vehicle bombs) and by the way they are initiated. IEDs are homemade and usually constructed for a specific target.

! Warning: This basic information should not be used in dealing with or dismantling an IED. Explosive ordnance disposal (EOD) technicians and local bomb squads are trained to accomplish this mission.

Descriptions

The design and placement of an IED is up to the imagination of the bomber. First and foremost, it is an object, regardless of its disguise, that is not supposed to be there. The best and most effective defense is to be aware of your surroundings. Based on your threat, if you think it does not belong in your area, consider it suspicious.

	*WARNING! If an object is considered suspicious, Do not touch it or move it . Evacuate the area and notify authorities. Any movement, however slight, may cause it to function.*

External Appearances of an IED

IEDs can be contained in almost anything. The item must be carried or driven to where it will placed, so concealment or masking of the device will be necessary. The outer container can be, but not limited to:
Pipe Bombs steel or PCV pipe section with end caps in nearly any configuration are the most prevalent type of containers.
Briefcase/Box/Back pack any style, color, or size; even as small as a cigarette pack.

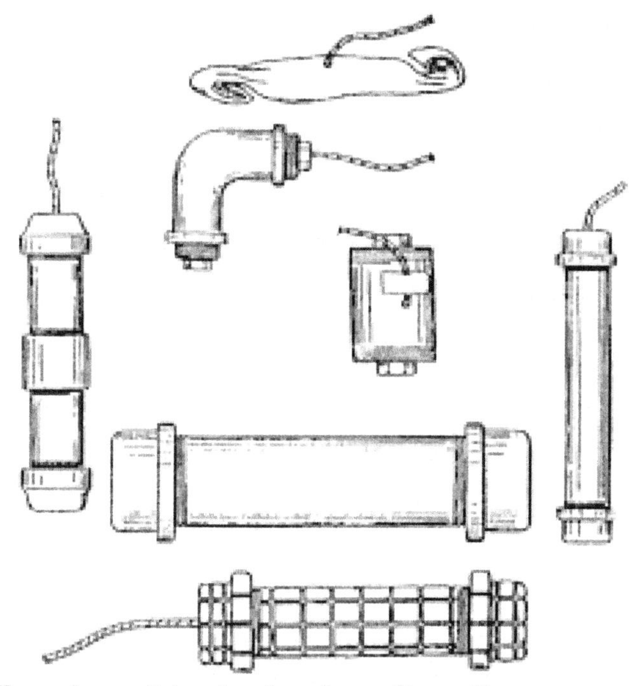

Sampling of possible pipe bomb configurations

Vehicle Bombs By far the most devastating (may contain thousands of pounds of explosives), vehicle bombs can be the easiest to conceal. Indicators may include inappropriate decals or an unfamiliar vehicle parked in your area. The device can be placed anywhere in the vehicle. A vehicle bomb is intended to create mass casualties or cause extensive property damage. Existing Objects Items that seem to have a purpose can be substituted or used as the bomb container. Some examples are fire extinguishers, propane bottles, trashcans, gasoline cans, or books.

Internally fused pipe bomb

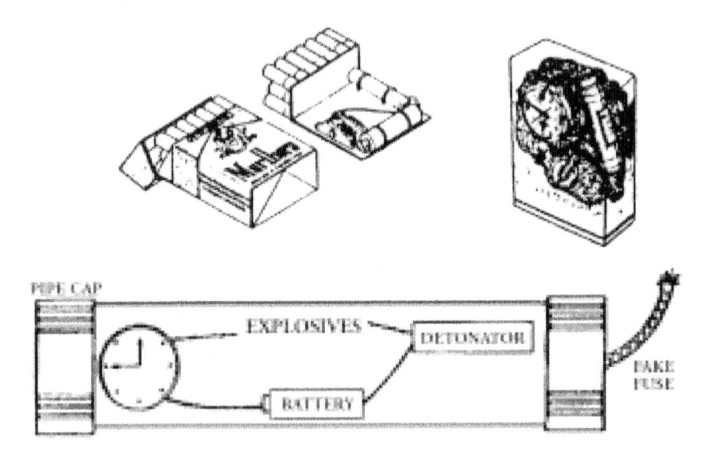

Internal Components
All devices require a firing train that consists of a fusing system, detonator, and main charge (explosive or incendiary). Any switch that can turn something on or off can be used to activate a device. Fusing systems can be categorized into the following:

- Time preset to detonate or arm the device at an unknown interval of time. The timer may be mechanical such as a kitchen timer, wind-up wristwatch, pocket watch, or electronic, i.e. digital wristwatches, integrated circuit chips, or solid-state timers.

Victim activated may be designed to function by pressure, pull, movement, vibration, tension release, or tilting the item. Booby-trapped is the best way to describe it.
Command sending a signal via radio frequency or through a hidden wire from a remote location, i.e. cell phone, walkie talkie, garage door remote etc.
Environmental designed to function when there is a change in temperature, pressure, light, sound, or magnetic field.

The detonator or blasting cap is a small explosive component, widely available from military and commercial sources, which can be initiated by a variety of mechanical and electrical devices. With the increased availability of blasting caps, fabrication and use of improvised detonators are on the decline. How-ever, the possibility of encountering one cannot be excluded.
Main charges can be used to burn, detonate or both, depending on the bombers desired effect. Explosives fall into three general categories.

Commercial Explosives used for property demolition, mining and blasting operations.

- Commercial explosives come in assorted shapes and consistencies including binary (two-part), slurries, gels, and standard dynamites.

Military Explosives differ from commercial explosives in several respects. Military explosives must have high rates of detonation, be relatively insensitive, and be usable underwater. TNT, C-4 plastic explosives, and military dynamite are some of the more common explosives associated with the military.
Improvised Explosives when manufactured explosives are not avail-able, it is relatively easy to obtain all of the ingredients necessary to make improvised explosives, such as ammonium nitrate (fertilizer), and potassium/sodium chlorate.
Incendiary improvised devices may be designed to burn. Included are some common materials used in incendiary devices: gasoline, iodine crystals, magnesium, glycerin, and aluminum powder.
Because of the vast variety of explosives and incendiary materials, any unknown solid, powder, crystal, or liquid should be treated with respect and not handled.

Where IEDs Can Be Placed

IEDs may be placed anywhere. A bomber wants to succeed without being caught. The level of security and the awareness of personnel will determine where and how an IED will be placed. Common areas where IEDs might be placed include:

- Outside areas: trash cans, dumpsters, mailboxes, bushes, storage areas, and parked vehicles.

Inside buildings: mail rooms, restrooms, trash cans, planters, inside desks or storage containers, false ceilings, utility closets, areas hidden by drapes or curtains, behind pictures, boiler rooms, under stairwells, recently repaired or patched segments of walls, floors, or ceilings, or in plain view.

What to do

In the event that a suspicious device is found, notify the proper authorities in accordance with existing bomb threat procedures. Security personnel should initiate and coordinate the evacuation in accordance with existing procedures, if necessary. Prior to their arrival, immediate actions should be taken.

- Using adequate cover (frontal and overhead) get as far away from the device as possible. Do not panic!

Keep away from glass windows that can become lethal fragmentation.
If the device is located outside the building, get low to the floor and go to the other side. Do not look out the window to see what is going on!
Increasing your distance from a suspicious device increases the chances of survival after a detonation.

BATF Explosive Standards

ATF	Vehicle Description	Maximum Explosives Capacity	Lethal Air Blast Range	Minimum Evacuation Distance	Falling Glass Hazard
	Compact Sedan	500 pounds 227 Kilos (In Trunk)	100 Feet 30 Meters	1,500 Feet 457 Meters	1,250 Feet 381 Meters
	Full Size Sedan	1,000 Pounds 455 Kilos (In Trunk)	125 Feet 38 Meters	1,750 Feet 534 Meters	1,750 Feet 534 Meters
	Passenger Van or Cargo Van	4,000 Pounds 1,818 Kilos	200 Feet 61 Meters	2,750 Feet 838 Meters	2,750 Feet 838 Meters
	Small Box Van (14 Ft. box)	10,000 Pounds 4,545 Kilos	300 Feet 91 Meters	3,750 Feet 1,143 Meters	3,750 Feet 1,143 Meters
	Box Van or Water/Fuel Truck	30,000 Pounds 13,636 Kilos	450 Feet 137 Meters	6,500 Feet 1,982 Meters	6,500 Feet 1,982 Meters
	Semi-Trailer	60,000 Pounds 27,273 Kilos	600 Feet 183 Meters	7,000 Feet 2,134 Meters	7,000 Feet 2,134 Meters

Vehicle Bombs
By far the most devastating (may contain thousands of pounds of explosives), vehicle bombs can be the easiest to conceal. Indicators may include inappropriate decals or an unfamiliar vehicle parked in your area. The device can be placed anywhere in the vehicle. A vehicle bomb is intended to create mass casualties or cause extensive property damage.

Indicative Behaviors of Suicide Bombers

There is no commonly accepted or developed profile of a suicide bomber. Studies indicate that the only characteristic accepted by experts is that the overwhelming majority are prepared to die in the service of their cause.

Most are 18-23, male, Islamic and single. But:

- Can be any race, color, sex

Can be older, married people

Nervousness, nervous glancing or other signs of being ill at ease. This may include sweating, "tunnel vision" (staring forward inappropriately) and repeated inappropriate prayer (e.g., outside the facility) or muttering. This may also include repeated entrances and exits from the building or facility.

Attempt to "Blend In" to environment. Might seem "Out of Place"

Inappropriate, oversized and/or loose-fitting clothes (e.g., a heavy overcoat on a warm day). Clothing gives impression that body is disproportionately larger than head or feet.

The appearance of carrying extra weight. Many bombs are packed with shrapnel such as ball bearings, screws, nails, nuts, bolts and screws that are blasted into the crowd upon detonation.

Profuse sweating that is inconsistent with weather conditions.

Bomber tries to avoid military, law enforcement.

No response to authoritative voice commands or direct salutation from a distance.

Deliberately walking toward a specific object or target, pushing their way through a crowd or going around barriers.

Constantly favoring one side or one area of the body as if wearing something unusual/uncomfortable (e.g., a bomb belt or vest). Pay attention to a person constantly adjusting waistbands or other clothing.

Bombers have been known to repeatedly pat themselves to verify that the bomb vest or belt is still attached.

Suspect may be carrying heavy luggage, bag, or wearing a backpack.

Keeping hands in pockets or cupping hands (as if holding a triggering device).

Newly shaved beards leaving unusual facial tan lines.

Scented anointing oil, which maybe obvious to someone in their vicinity.

Behavior is consistent with no future, e.g. individual purchases a one-way ticket or is unconcerned about receipts for purchases, or receiving change.

CIVIL UNREST / RIOTS

Disclaimer: this information is advisory in nature and is not
Avoid potential riot areas if at all possible. If you know an area is viable for a potential riot but you can't avoid traveling there, take some simple precautions to help protect yourself. First, be prepared for the worst; the unexpected can happen at any moment. Crowds are dangerous when they're in a revolting temperament and normally placid people can turn aggressive just by being in the presence of other aggressive people.

Keep up-to-date
Riots don't drop out of thin air. Generally, there may be signs of public anger and violence at least one day (in some cases even 3-4 days) before the actual riot. Reading the newspapers and following the news may give you a warning about impending protests, rallies, marches etc. Being informed and avoiding troubled areas may be your best defense.

Avoid Main Thoroughfares Riots tend to occur in public squares, at government buildings and on main thoroughfares. Take alleys and side streets away from the crowd.

Wear clothes that minimize the amount of exposed skin (long pants and long-sleeved shirts) when going out. Do not wear clothing that could be interpreted as military or police wear in any way; avoid wearing anything that looks like a uniform. Try and dress neutrally.

Avoid confrontation by keeping your head down. Walk at all times. If you run or move too quickly, you might attract unwanted attention. If a riot or instance of civil unrest does occur: Get inside and stay inside. Typically riots occur in the streets or elsewhere outside. Being inside, especially in a large, sturdy structure can be your best protection to endure the storm such as a basement.

Try to find at least two possible exits in case you need to evacuate the building in a hurry. Keep doors and windows locked, avoid watching the riot from windows or balconies, and try to move to inside rooms, where the danger of being hit by stones or bullets is minimized.

Other concern areas & useful information: If a riot breaks out in a stadium, your response should be different depending on where you are in relation to the rioters. If you are in the midst of a riot, you should try to quickly move to an exit. Don't run, however, and try not to jostle others. If you are at some distance from the action, stay where you are unless instructed to move by police or security personnel. Don't rush for the exits unless you're in imminent danger. People are frequently trampled by stampeding crowds near exits.

Watch your footing in a mob situation. If you stumble and fall to the ground you're likely to be trampled. If you fall down, pull yourself up into a ball. Protect your face, ears and internal organs. In this position you are a smaller object that can be avoided. You will receive less damage if you are stepped on.

Take some supplies Have some emergency supplies on hand such as water, flashlight, snacks, map and whatever else might help you survive until things blow over.

ACTIVE SHOOTER

! Disclaimer: this information is advisory in nature and is not intended to identify all scenarios or situations a person might encounter.
! Following these guidelines will **not** guarantee your safety.

Note: This information should be read and understood immediately. During a shooting is no time to try to figure out what to do. Regular review and practice will help you prepare for an emergency.

Profile of an Active Shooter

An Active Shooter is an individual actively engaged in killing or attempting to kill people in a confined and populated area; in most cases, active shooters use firearms(s) and there is no pattern or method to their selection of victims. Active shooter situations are unpredictable and evolve quickly. Typically, the immediate deployment of law enforcement is required to stop the shooting and mitigate harm to victims. Because active shooter situations are often over within 10 to 15 minutes, before law enforcement arrives on the scene, individuals must be prepared both mentally and physically to deal with an active shooter situation.

Personal Response

Good practices for coping with an active shooter situation:

- Be aware of your environment and any possible dangers
- Take note of the two nearest exits in any facility you visit
- If you are in an office, stay there and secure the door
- If you are in a hallway, get into a room and secure the door
- As a last resort, attempt to take the active shooter down. When the shooter is at close range and you cannot flee, your chance of survival is much greater if you try to incapacitate him/her.
- Call 911 when it is safe to do so!

Business Response

Quickly determine the most reasonable way to protect your own life. Remember that customers and clients are likely to follow the lead of employees and managers during an active shooter situation.

Evacuate

If there is an accessible escape path, attempt to evacuate the premises. Be sure to:

- Have an escape route and plan in mind
- Evacuate regardless of whether others agree to follow
- Leave your belongings behind

- Help others escape, if possible
- Prevent individuals from entering an area where the active shooter may be
- Keep your hands visible
- Follow the instructions of any police officers
- Do not attempt to move wounded people
- Call 911 when you are safe

Hide out

If evacuation is not possible, find a place to hide where the active shooter is less likely to find you.

Your hiding place should:

- Be out of the active shooter's view
- Provide protection if shots are fired in your direction (i.e., an office with a closed and locked door)
- Not trap you or restrict your options for movement

To prevent an active shooter from entering your hiding place:

- Lock the door
- Blockade the door with heavy furniture

If the active shooter is nearby:

- Lock the door
- Silence your cell phone and/or pager
- Turn off any source of noise (i.e., radios, televisions)
- Hide behind large items (i.e., cabinets, desks)
- Remain quiet

If evacuation and hiding out are not possible:

- Remain calm
- Dial 911, if possible, to alert police to the active shooter's location
- If you cannot speak, leave the line open and allow the dispatcher to listen

Take action against the active shooter

As a last resort, and only when your life is in imminent danger attempt to disrupt and/or incapacitate the active shooter by:

- Acting as aggressively as possible against him/her
- Throwing items and improvising weapons
- Yelling
- Committing to your actions

Working with law enforcement

- Law enforcement's purpose is to stop the active shooter as soon as possible. Officers will proceed directly to the area in which the last shots were heard.
- Officers usually arrive in teams of four (4)
- Officers may wear regular patrol uniforms or external

bulletproof vests, Kevlar helmets, and other tactical equipment
- Officers may be armed with rifles, shotguns, handguns
- Officers may use pepper spray or tear gas to control the situation
- Officers may shout commands, and may push individuals to the ground for their safety

When law enforcement arrives:
- Remain calm, and follow officers' instructions
- Put down any items in your hands (i.e., bags, jackets)
- Immediately raise hands and spread fingers
- Keep hands visible at all times
- Avoid making quick movements toward officers such as attempting to hold on to them for safety
- Avoid pointing, screaming and/or yelling
- Do not stop to ask officers for help or direction when evacuating, just proceed in the direction from which officers are entering the premises

Information to provide to law enforcement or 911 operator:
- Location of the active shooter
- Number of shooters, if more than one
- Physical description of shooter/s
- Number and type of weapons held by the shooter/s
- Number of potential victims at the location

The first officers to arrive to the scene will not stop to help injured persons. Expect rescue teams comprised of additional officers and emergency medical personnel to follow the initial officers. These rescue teams will treat and remove any injured persons. They may also call upon able-bodied individuals to assist in removing the wounded from the premises.

Once you have reached a safe location or an assembly point, you will likely be held in that area by law enforcement until the situation is under control, and all witnesses have been identified and questioned. Do not leave the safe location or assembly point until law enforcement authorities have instructed you to do so.

TRAVEL SAFETY

HOTEL SAFETY & SECURITY

! *Disclaimer: this information is advisory in nature and is not intended to identify all scenarios or situations a person might encounter.*

! *Following these guidelines will not guarantee your safety.*

Hotel Safety & Security

- Avoid taking a street level room. Choose the second or third floors which are usually too high for easy outside access yet low enough to be reached by fire equipment.
- Use elevators rather than stairwells. Stand near the control panel so if you are threatened, you can push the alarm button.
- Locate exits within the hotel and develop a plan in case of fire or other emergency.
- Report lost keys immediately and consider changing rooms.
- When in the hotel room, secure the door and windows and keep them locked.
- When you leave your room, put the "Do Not Disturb" sign on your door; do not leave indicators showing that you are out. In fact, leave the television or radio on, giving the impression that the room is occupied.
- Do not leave anything of value (money, tickets, camera, etc.) or work-related items (briefcases, computers, etc.) in your room when you go out, even if it is locked in your suitcase.
- Do not accept deliveries to your room unless previously arranged and you are certain of the source and contents.
- Be sure of the identity of visitors before opening the door of your hotel room. Don't meet strangers at your hotel room, or at unknown or remote locations.
- Keep your room key with you instead of leaving it at the front desk.
- Don't advertise to others when you are out of your room. For example, request that housekeeping make up your room while you are at breakfast, rather than leaving a "Please Service This Room" sign on the door knob.

Hotel Key Security

When you check out of a hotel that uses swipe cards for keys, do NOT turn them in. Destroy them. They contain your address, phone number and credit card numbers. Someone with a card reader can easily access the information.

Hotel fires

You must aggressively take responsibility for the safety of yourself and your family. Think "contingency plan" and discuss it with your dependents. Begin planning your escape from a fire as soon as you check into a hotel. When a fire occurs, you can then act without panic and without wasting time.

- Request a lower floor, ideally the second or third. Selecting a room no higher than the second floor enables you to jump to safety. Although most fire departments can reach above the second floor, they may not get to you in time or position a fire truck on your side of the building.
- Locate exits and stairways as soon as you check in; be sure the doors open. Count the number of doors between your room and exit or stairway. In a smoke-filled hallway, you may have to "feel" your way to an exit. Form a mental map of your escape route.
- If the hotel has a fire alarm system, find the nearest alarm. Be sure you know how to use it. You may have to activate it in the dark or in dense smoke.
- Ensure that your room windows open and that you know how the latches work. Look out the window and mentally rehearse your escape through it. Make note of any ledges or decks that will aid escape.
- Check the smoke detector by pushing the test button. If it does not work, have it fixed or move to another room. Better yet, carry your own portable smoke detector (with the battery removed while traveling). Place it in your room by the hall door near the ceiling.
- Keep the room key and a flashlight on the bedside table so that you may locate the key quickly if you have to leave your room.

If a Fire Starts

- If you awake to find smoke in your room, grab your key and crawl to the door on your hands and knees. Do not stand. Smoke and deadly gases rise while the fresher air will be near the floor.
- Before you open the door, feel it with the palm of your hand. If the door or knob is hot, the fire may be right outside. Open the door slowly. Be ready to slam it shut if the fire is close by.
- If your exit path is clear, crawl into the hallway. Be sure to close the door behind you to keep smoke out in case you have to return to your room. Take your key, as most hotel doors lock automatically. Stay close to the wall to avoid being trampled.
- Do not use elevators during a fire. They may malfunction, or if they have heat-activated call buttons, they may take you

directly to the fire floor.

- As you make your way to the fire exit, stay on the same side as the exit door. Count the doors to the exit.
- When you reach the exit, walk down the stairs to the first floor. Hold onto the handrail for guidance and protection from being knocked down by other occupants.
- If you encounter heavy smoke in the stairwell, do not try to run through it. You may not make it. Instead, turn around and walk up to the roof fire exit. Prop the door open to ventilate the stairwell and to keep from being locked out. Find the windward side of the roof, sit down, and wait for firefighters to find you.
- If all exits are blocked or if there is heavy smoke in the hallway, you will be better off staying in your room. If there is smoke in your room, open a window and turn on the bathroom vent.
- Do not break the window unless it cannot be opened. You might want to close the window later to keep smoke out, and broken glass could injure you or people below.
- If your phone works, call the desk to tell someone where you are, or call the fire department to report your location in the building. Hang a bed sheet out the window as a signal.
- Fill the bathtub with water to use for firefighting. Bail water onto your door or any hot walls with an ice bucket or waste basket. Stuff wet towels into cracks under and around doors where smoke can enter. Tie a wet towel over your mouth and nose to help filter out smoke. If there is fire outside your window, take down the drapes and move everything combustible away from the window.
- If you are above the second floor, you probably will be better off fighting the fire in your room than jumping. A jump from above the third floor may result in severe injury or death.
- Remember that panic and a fire's by-products, such as super-heated gases and smoke, present a greater danger than the fire itself. If you know your plan of escape in advance, you will be less likely to panic and more likely to survive

Identity Theft / Hotel Key Security Precautions

- Photocopy both the front and back of the entire contents of your wallet - credit cards, licenses, etc. and keep the copies in a safe and secure location. This will enable you to cancel your credit card as soon as possible if is lost or stolen.
- Do not sign the cards. Instead, put "Photo Id Required"
- Carry credit cards separately if possible. Carry only the minimum number of credit cards needed.
- When you write a check, never allow the salesperson to write down your credit card number on the check. If paying by

credit card, never let the salesperson write down your driver's license number.

- Avoid signing a blank receipt, whenever possible. Draw a line through blank spaces above the when you sign card receipts.
- When you use a credit card to make a purchase, maintain visual contact with the card and make sure no extra imprints of your card are made to other charge slips.

- It is a good idea to retain your credit receipts and check them against the monthly billing statement.
- In the event your credit card is lost or stolen.
- Immediately notify the credit card company. Most issuing banks or companies can be reached 24 hours a day, 365 days a year. The majority of fraudulent purchases are made within 48 hours of the loss.
- Credit card thieves may sometimes call the victim, inform the person that their credit card has been found and that it is being returned. This ploy gives the thief time to go on a charging spree because the card holder never calls to cancel the card.
- File a police report immediately in the jurisdiction where your card was stolen.
- Important! Call the three national credit reporting organizations immediately and put a fraud alert on your name and SS number.
- By virtue of the Fair Credit Billing Act (FCBA), if you report the loss of a credit card before it is used, the card issuer cannot hold you responsible for any unauthorized charges. If a thief uses your credit card before you report it missing, the most you will owe for unauthorized charges on each credit card is $50.00.
- Report credit card fraud to any one of the three major credit reporting bureaus, and they will contact the other two credit bureaus for you.

Credit bureaus

Experian

To Report Fraud:
888-397-3742
PO Box 9530
Allen, TX 75013
TDD 1-800-972-0322

To Order Your Credit Report
888-397-3742
PO Box 2002
Allen, TX 75013

Equifax

To Report Fraud:
800-525-6285
PO Box 740241
Atlanta, GA 30374-0241
TDD 1-800-255-0056

To Order Your Credit Report
800-685-1111
PO Box 740241
Atlanta, GA 30374-0241

Transunion

To Report Fraud:
800-680-7289
PO Box 6790
Fullerton, CA 92634
TDD 1-877-553-7803

To Order Your Credit Report
800-888-4213
PO Box 1000
Chester, PA 19022

NOTES:

Check Security

- When writing a check to pay on your credit card accounts, do not use the complete number instead, and use only the last four digits - the card company knows the rest.
- Never give out your Social Security number.
- Put your initials on your checks instead of your full name.
- Put your work phone on the checks instead of your home number
- Use a PO box (if possible) instead of a home address (to prevent burglary)
- Don't print or offer any more information than is necessary
- If you've had checks stolen or bank accounts set up fraudulently in your name, call these check guarantee companies: Telecheck at 800-366-2425; and the National Processing Company at 800-526-5380. They can flag your file so that counterfeit checks will be refused.

If your social security number was used fraudulently, report the problem to the Social Security Administration's Fraud Hotline at 800-269-0271. In extreme cases of fraud, it may be possible for you to get a new social security number. Federal trade commission identity theft complaint hotline at 1-877-id theft (1-877-438-4338) or online at www.consumer.gov/idtheft

CONTACTS

Federal Agencies

Pipeline and Hazardous Materials Safety Administration
U.S. Department of Transportation 1200 New Jersey Ave, SE., Washington, DC 20590 Hazardous Materials Info-Line: 800-467-4922

Publications and Reports
Fax: 202-366-7342; Telephone: 202-366-4900 E-Mail: training@dot.gov http://hazmat.dot.gov

Federal Aviation Administration
U.S. Department of Transportation 800 Independence Avenue, SW., Washington, DC 20591 Telephone: 1-866-TELL-FAA (1-866-835-5322)
http://www.faa.gov

Federal Motor Carrier Safety Administration
U.S. Department of Transportation 1200 New Jersey Ave, SE., Washington, DC 20590 Telephone: 800-832-5660
http://www.fmcsa.dot.gov

Federal Railroad Administration
U.S. Department of Transportation 1200 New Jersey Ave, SE., Washington, DC 20590 Telephone: 202-493-6024
http://www.fra.dot.gov

Bureau of Alcohol, Tobacco, Firearms and Explosives
Explosives Industry Programs Branch 99 New York Avenue, NE, Room 6N-672 Washington, DC 20226 202-648-7120 E-Mail: EIPB@atf.gov http://www.atf.gov/

Bureau of Alcohol, Tobacco, Firearms and Explosives
U.S. Bomb Data Center 99 New York Avenue, NE, Room 8S-295 Washington, DC 20226 800-461-8841 E-Mail: USBDC@atf.gov
http://www.atf.gov/

Transportation Security Administration
601 South 12th Street Arlington, VA 20598 Telephone: 866-289-9673
http://www.tsa.gov
United States Coast Guard
2100 Second Street, SW., STOP 7000 Washington, DC 20593 Telephone: 202-493-1713
http://www.uscg.mil

Industry Associations/Organizations

American Chemistry Council
700 Second Street, NE. Washington, DC 20002 Telephone: 202-249-7000
http://www.americanchemistry.com
American Petroleum Institute
Washington, DC 20005 Telephone: 202-682-8000 1220 L Street, NW.
http://www.api.org
American Society for Industrial Security
1625 Prince Street Alexandria, VA, 22314 Telephone: 703-519-6200
http://www.asisonline.org
American Trucking Association
950 North Glebe Road, Suite 210 Arlington, VA 22203 Telephone: 703-838-1700
http://www.truckline.com

Association of American Railroads
425 Third Street, SW. Washington, DC 20024 Telephone: 202-639-2100
http://www.aar.org
Center for Chemical Process Safety American Institute of Chemical Engineers
3 Park Avenue New York, N.Y. 10016-5991 Telephone: 212-591-7319
http://www.aiche.org/ccp
Chlorine Institute
1300 Wilson Blvd, Suite 525 Arlington, VA 22209 Telephone: 703-894-4140
http://www.chlorineinstitute.org
Compressed Gas Association
4221 Walney Road, 5th Floor Chantilly, VA 20151 Telephone: 703-788-2700
http://www.cganet.com
The Fertilizer Institute
425 Third Street SW, Suite 950 Washington, DC 20024 Telephone: 202-962-0490
http://www.tfi.org
Institute of Makers of Explosives
1120 19th Street, Suite 310, NW. Washington, DC 20036 Telephone: 202-429-9280
http://www.ime.org
National Association of Chemical Distributors
1555 Wilson Blvd, Suite 700 Arlington, VA 22209 Telephone: 703-527-6223
http://www.nacd.com
ENHANCED SECURITY REQUIREMENTS **16**

National Propane Gas Association
1899 L Street NW, Suite 350, Washington, DC 20036 Teléfono: 202-466-7200
http://www.npga.org
National Tank Truck Carriers
950 North Glebe Road, Suite #520 Arlington, Virginia 22203-4183 Telephone: 703-838-1960
http://www.tanktransport.com
Security Industry Association
635 Slaters Lane Alexandria, Virginia 22314 Telephone: 866-817-8888
http://www.siaonline.org
Synthetic Organic Chemical Manufacturers Association
1850 M Street, NW, Suite 700 Washington, DC 20036 Telephone: 202-721-4100
http://www.socma.com
Additional Security Requirement Resources
TSA Security Requirements
http://www.tsa.gov/travelers/airtravel/acceptable_documents.shtm
http://www.tsa.gov/assets/pdf/cargo_final_rule_5-26-06.pdf
NRC Security Requirements:
http://www.nrc.gov/security/byproduct/orders.html
NNSA Security:
http://nnsa.energy.gov/
PHMSA Security:
http://www.phmsa.dot.gov/hazmat/security
USCG Facility Requirements:
http://www.uscg.mil/hq/cg5/cg522/cg5222/

2014

January

S	M	T	W	T	F	S
			1	2	3	4
5	6	7	8	9	10	11
12	13	14	15	16	17	18
19	20	21	22	23	24	25
26	27	28	29	30	31	

February

S	M	T	W	T	F	S
						1
2	3	4	5	6	7	8
9	10	11	12	13	14	15
16	17	18	19	20	21	22
23	24	25	26	27	28	

March

S	M	T	W	T	F	S
						1
2	3	4	5	6	7	8
9	10	11	12	13	14	15
16	17	18	19	20	21	22
23	24	25	26	27	28	29
30	31					

April

S	M	T	W	T	F	S
		1	2	3	4	5
6	7	8	9	10	11	12
13	14	15	16	17	18	19
20	21	22	23	24	25	26
27	28	29	30			

May

S	M	T	W	T	F	S
				1	2	3
4	5	6	7	8	9	10
11	12	13	14	15	16	17
18	19	20	21	22	23	24
25	26	27	28	29	30	31

June

S	M	T	W	T	F	S
1	2	3	4	5	6	7
8	9	10	11	12	13	14
15	16	17	18	19	20	21
22	23	24	25	26	27	28
29	30					

July

S	M	T	W	T	F	S
		1	2	3	4	5
6	7	8	9	10	11	12
13	14	15	16	17	18	19
20	21	22	23	24	25	26
27	28	29	30	31		

August

S	M	T	W	T	F	S
					1	2
3	4	5	6	7	8	9
10	11	12	13	14	15	16
17	18	19	20	21	22	23
24	25	26	27	28	29	30
31						

September

S	M	T	W	T	F	S
	1	2	3	4	5	6
7	8	9	10	11	12	13
14	15	16	17	18	19	20
21	22	23	24	25	26	27
28	29	30				

October

S	M	T	W	T	F	S
			1	2	3	4
5	6	7	8	9	10	11
12	13	14	15	16	17	18
19	20	21	22	23	24	25
26	27	28	29	30	31	

November

S	M	T	W	T	F	S
						1
2	3	4	5	6	7	8
9	10	11	12	13	14	15
16	17	18	19	20	21	22
23	24	25	26	27	28	29
30						

December

S	M	T	W	T	F	S
	1	2	3	4	5	6
7	8	9	10	11	12	13
14	15	16	17	18	19	20
21	22	23	24	25	26	27
28	29	30	31			

2015

January

S	M	T	W	T	F	S
				1	2	3
4	5	6	7	8	9	10
11	12	13	14	15	16	17
18	19	20	21	22	23	24
25	26	27	28	29	30	31

February

S	M	T	W	T	F	S
1	2	3	4	5	6	7
8	9	10	11	12	13	14
15	16	17	18	19	20	21
22	23	24	25	26	27	28

March

S	M	T	W	T	F	S
1	2	3	4	5	6	7
8	9	10	11	12	13	14
15	16	17	18	19	20	21
22	23	24	25	26	27	28
29	30	31				

April

S	M	T	W	T	F	S
			1	2	3	4
5	6	7	8	9	10	11
12	13	14	15	16	17	18
19	20	21	22	23	24	25
26	27	28	29	30		

May

S	M	T	W	T	F	S
					1	2
3	4	5	6	7	8	9
10	11	12	13	14	15	16
17	18	19	20	21	22	23
24	25	26	27	28	29	30
31						

June

S	M	T	W	T	F	S
	1	2	3	4	5	6
7	8	9	10	11	12	13
14	15	16	17	18	19	20
21	22	23	24	25	26	27
28	29	30				

July

S	M	T	W	T	F	S
			1	2	3	4
5	6	7	8	9	10	11
12	13	14	15	16	17	18
19	20	21	22	23	24	25
26	27	28	29	30	31	

August

S	M	T	W	T	F	S
						1
2	3	4	5	6	7	8
9	10	11	12	13	14	15
16	17	18	19	20	21	22
23	24	25	26	27	28	29
30	31					

September

S	M	T	W	T	F	S
		1	2	3	4	5
6	7	8	9	10	11	12
13	14	15	16	17	18	19
20	21	22	23	24	25	26
27	28	29	30			

October

S	M	T	W	T	F	S
				1	2	3
4	5	6	7	8	9	10
11	12	13	14	15	16	17
18	19	20	21	22	23	24
25	26	27	28	29	30	31

November

S	M	T	W	T	F	S
1	2	3	4	5	6	7
8	9	10	11	12	13	14
15	16	17	18	19	20	21
22	23	24	25	26	27	28
29	30					

December

S	M	T	W	T	F	S
		1	2	3	4	5
6	7	8	9	10	11	12
13	14	15	16	17	18	19
20	21	22	23	24	25	26
27	28	29	30	31		

INDEX

CHARTS / MAPS

CPSIA information can be obtained at www.ICGtesting.com
Printed in the USA
BVOW08s0029060415

394844BV00021B/191/P

9 780692 210680